To Rosie
    On the occasion
    Of your first Show.

            Uncle.

                    August 1978

The Wild Goats of Great Britain and Ireland

The
WILD GOATS
of Great Britain
and Ireland

G. KENNETH WHITEHEAD

DAVID & CHARLES
Newton Abbot

ISBN 0 7153 5508 2

Set in 11 on 13pt Times New Roman
by C. E. Dawkins (Typesetters) Limited
and printed in Great Britain
by Redwood Press Limited Trowbridge and London
for David & Charles (Publishers) Limited
South Devon House Newton Abbot Devon

Contents

Illustrations

Photographs not otherwise acknowledged are by the author

Introduction

In parts of the British Isles, especially on the western seaboard of Scotland and adjacent islands, small herds of goats seek a scanty living on the cliffs and rocky mountain tops, to form an interesting addition to the fauna of the islands. Yet these beasts are not derived from wild stock, for their race has sprung up from imported and semi-domesticated goats which many years ago escaped or were deliberately liberated. Many generations of almost unmolested freedom have produced a hardy race which is extremely shy and dubious of human intrusion. Indeed 'these mountain goats are a solemn set' as Tristram wrote, 'and by the gravity of their demeanour excite a suspicion that they have no youth, and never were kids'. Although, therefore, there are no truly wild goats in Britain today, there are many *feral* goats, some of which have been living a free and undomesticated existence for hundreds of years. These goats, particularly when persecuted, have shown themselves to be just as elusive as wild deer, and in many cases, therefore, can quite accurately be described as wild—and this has been their general description in the book.

Apart from two articles on wild goats by A. D. Buchanan Smith (1932) and Hugh Boyd Watt (1937), both of whom dealt with Scotland, and a number of short articles or letters in such papers as *The Field, Country Life* and *The Shooting Times and Country Magazine,* information concerning the wild goats of Britain is sadly lacking.

Several books have, of course, been written on domestic goat breeding—the most important being Holmes Pegler's *The Book of the Goat* and David Mackenzie's *Goat Husbandry*—but mention of the wild animal is dismissed in a paragraph or so. And there are also such writers as J. G. Millais and H. Frank Wallace who have included an odd chapter on this animal in some of their sporting literature, but their interest in the feral goat was in its chase.

There was, as the Reverend W. B. Daniel observed in *Rural Sports* (1813) 'a peculiar kind of language invented by Sportsmen of the Middle Ages which it was necessary to be acquainted with'. C. E. Hare in *The Language of Field Sports* has recorded many of these ancient terms. For instance, *trip, tryppe* or *trippe*—which is probably the correct term—was commonly used to denote a herd of goats, as was also *tribe*. Indeed, Hare suggested that *trip* or *tripp* may be a corruption of tribe, which rendering is given by Strutt in *The Sports and Pastimes of the People of England*, (1801). Other collective names used for goats are flock or herd, formerly spelt heard. At rutting time, when the male goat makes a sound, he was said to *rattle*.

There is a great variety in the spelling of the word for goats, such as *gete, geet, geete, geates, gayte, gait, goete, gotes* and *goates*. It is suggested that *geet* is the plural of goat just as teeth is the plural of tooth. Tubervile (1576) called the males goats and the females geats. Gat seems to have been the Anglo-Saxon word for goat. The modern names for the male and female goat have generally been billy and nanny respectively; Pegler, however, thought these names vulgar, and suggested that it was high time these childish terms should be abandoned in favour of the more sensible English words of buck and doe.

In America and most of the colonies the majority of goat keepers do refer to their goats as bucks and does but I am informed by the Secretary of the British Goat Society that their members are very conservative, referring to their animals mainly as 'goats'. When it is necessary to differentiate between the sexes—say at shows—the animal is simply referred to as 'male goat' or 'female goat', whilst 'stud male' is used for the working male.

Other terms used for male and female goat have been 'he-goat' and 'she-goat', particularly the latter, whilst for the young, kid has been long preserved. A young male goat is often referred to as a 'buckling'. In south-west Scotland I have heard the word *heverin* used to denote a castrated billy goat, whilst in Wales the male has been called *bwch,* the female *gafr* and the kid *mynn.*

Other Gaelic words that have been used in connection with the goat include *aibhreann* (castrated); *habrun* or *haburn* (three-year-old castrated); *bean* (milker); *boc* (male); *ealt-ghobhar* (trip of); *gabhar fhiadhain* (wild goat); *gabhrag* (flock of); *meann* (kid) and *meann-bhoc* (buck kid).

In this book, the male and female goat have generally been referred to as billy and nanny respectively, whilst herd, in preference to trip, has been used to denote a collection of animals.

1
Wild Goats of the World

Goats belong to the order *Artiodactyla,* which comprises the hoofed mammals having an even number of toes, within which they form part of the large family *Bovidae.* This family includes all the cattle, sheep, goats, deer and antelopes and hence many of the most important domesticated animals, the goats forming the genus *Capra.* The distribution of goats as indigenous animals is not quite world-wide since they are absent from the Americas and Australasia, but throughout the world domestic goats have gone wild in suitable localities, being especially plentiful, for example, in the mountains of New Zealand.

It is true, of course, that in western North America there is a so-called Rocky Mountain Goat, but this beautiful white, shaggy-haired creature is not a true goat and belongs to a different genus, *Oreamnos.* Its range includes the Yukon, Alberta and British Columbia in Canada, extending in the south to Washington and in the west to Alaska.

The various species of *Capra,* which include the true goats, ibexes and markhors, are distributed over the mountainous districts of southern Europe, the islands of the Grecian archipelago and north-eastern Africa, whence they extend eastwards through the Caucasus, Asia Minor, Arabia, Iran, Sind and Baluchistan to central Asia. One of the ancestors of the domestic goat *Capra hircus hircus* is undoubtedly the

13

wild subspecies *Capra hircus aegagrus* whose range includes the Caucasus, some of the Greek islands, Asia Minor and Iran. The species is also known as the Grecian or Persian ibex and, in Iran, as the *Pasang,* meaning 'rock-footed'— an allusion to its skill in rock climbing. The relationships of many of the populations of this animal, especially in the Greek islands and other localities where it has often interbred with feral domestic goats, are confused. Another subspecies is *C.h.blythi*—sometimes known as the Sind ibex—which inhabits western Sind, Baluchistan and Turkmenia.

In build, this goat is relatively slender, with a shoulder height which seldom exceeds 37in, the general colour in winter is a brownish grey which becomes yellowish or redder in summer; the inner sides of the thighs and underparts, are lighter coloured or white.

The horns of the adult male are triangular in cross-section and have an anterior edge curved like a scimitar and laterally compressed so as to form a sharp leading edge on which irregular notches develop with age. A good pair of male horns may measure 50in, but those of the female are very much shorter and not keeled.

The true ibexes have a wide distribution, ranging from Spain in the west, through some of the Mediterranean countries of southern Europe, Asia Minor, northern India, central Siberia to Mongolia in the east. Ibex are mountain-dwelling animals mostly living in small herds and climbing with great agility; their horns are curved like those of *Capra hircus aegagrus* but instead of a sharp leading edge they have a relatively flat anterior surface broken by a number of horizontal ridges.

In western Europe two species are represented, *Capra pyrenaica* in Spain and *C.ibex* in the alpine regions of northern Italy, Switzerland, Austria, Yugoslavia and France. Both species, at the beginning of the century were near extinction but fortunately their plight was recognised and strict preservation measures ensured their survival.

In 1905 it was estimated that the population of Spanish ibex at Gredos in central Spain was two males and five females. At about this time the area was handed over to King Alfonso XIII as a royal hunting preserve and the Marquis of Villaviciosa was commissioned to preserve the ibex. Success was immediate and within some twelve years the ibex had increased to about 400. In 1932 Gredos became a National reserve under the control of the Spanish State Tourist Department and the ibex have increased steadily; they now number several thousand and some are permitted to be shot under licence each year.

Although the majority of Spanish ibex, of which three races are recognised, is to be found in the reserve at Gredos (*C.p.victoriae*) small numbers may be met with in the Sierra Nevada (*C.p.hispanica*) as well as in such isolated districts as the Sierra de Cazorla, Morena and Serrania de Ronda in the south, the Montes de Tortosa in the east and in the Pyrenees (*C.p.pyrenaica*), where a few years ago they were thought to be extinct. In the last locality its present status is unknown, but it is doubtful if there are more than forty animals. A fourth race, the Portuguese ibex (*C.p. lusitanica*), became extinct at the end of the last century.

The Spanish ibex is smaller than the alpine ibex, an average male standing 30in at the shoulder; it is also lighter in colour, the general body colour being pale brown with the outer sides of the limbs black, a black band on the lower parts of the flanks and a short, black mane which continues narrowly along the back. The forehead is blackish and the short beard is also dark in colour. The horns are more like those of the Caucasian tur than those of the alpine ibex and for this reason the animal was formerly known as the Spanish tur.

The vicissitudes of the alpine ibex *Capra ibex,* often referred to as the steinbock, follow a similar pattern to the Spanish species. Formerly generally distributed throughout the higher Alps of Italy, Switzerland, France, Bavaria and Austria it had, long before the last century, become extinct everywhere except for a few hundred in the Gran Paradiso of Italy. The young Victor Emmanuel, subsequently to become Victor Emmanuel II, saw the precarious position that the animal had reached and published a decree entirely forbidding all hunting of it. Helped by Joseph Zumstein, the inspector of forests, he conceived the idea of turning the massive range of mountains known as the Gran Paradiso into a royal hunting ground, and in 1836 his plans were put into effect.

For the next ninety years the Gran Paradiso remained a royal preserve and by 1914 the ibex were estimated at about 6,000 head, this being reduced during the war years to about 4,000. After World War I the king, Victor Emanuel III, proposed relinquishing the Gran Paradiso preserve so that it could be made into a national park. This suggestion was not well received by the local inhabitants and for a time shooting in the area seems to have gone on unrestricted, with the result that the ibex were soon reduced to about 2,000 animals. However, in 1922 the Gran Paradiso was formally declared a national park and for the next eight years no hunting was permitted, allowing the ibex to increase to about 4,000 by 1930 when a limited number were allowed to be shot under licence.

When the Facists took over in Italy the park came under military control and ibex stocks began to dwindle; by 1940 the population stood at about 1,800 and a few years later, after the war, only some 300-400 remained. Fortunately immediate steps were taken for their preservation and in 1945 all hunting was forbidden; for the third time within the space of little more than a century the species had received a reprieve. Today its population in Italy alone exceeds 2,000, whilst in neighbouring Switzerland, as a result of introductions, close on 3,000 ibex are living in a number of widely separated colonies that have been established in the various cantons.

In Switzerland the ibex had vanished long before the last century, indeed, as long ago as 1638 its position had become so precarious that the death penalty was decreed for anyone convicted of killing one. One of the reasons for this state of affairs was that the horns were then much used for making goblets, as it was claimed that the presence of certain poisons was betrayed by a cup of ibex horn. Shavings of the horn were also believed a cure for hysterics and the blood good 'against the stone'; small wonder then that these superstitions gave the ibex such a high value and that there were always men ready to risk their lives to hunt them.

One of the most successful of the Swiss introductions is the herd started by Andreas Rauch, who more than anyone else is associated with the re-introduction of ibex to Switzerland. About 1921 two female ibex appeared on the Piz Albris, where Rauch was game warden, having wandered down from the Parc National where the species had been introduced the previous year. He arranged for two males to be brought in from the Peter and Paul Game Park at St Gallen and from this beginning sprang a herd which now numbers over 500 head. Andreas Rauch died in 1942 and now lies in the tiny graveyard at the foot of the Piz Albris slopes, his tombstone suitably inscribed in commemoration of his work in re-establishing the ibex in Switzerland.

The adult male alpine ibex stands about 33in at the shoulder and the horns may measure 3ft in length, with exceptional specimens exceeding 40in. The general body colour is a dusky grey-brown and the males carry a short beard.

17

The rutting season is the latter part of November and early December, and as gestation lasts about six months the young are dropped near midsummer. It is understood that the females are in season for about twenty-two hours and it is estimated that of the total of females of breeding age only about 18 per cent would bear a kid every year. Over a period of two years, however, possibly 80 per cent would breed, perhaps 2 per cent bearing twins. The mortality among ibex is considerable and in some years when the winter has been severe upwards of twenty or more have perished in a single avalanche, a danger that even the sagacity of an old male cannot avoid.

The European ibex belong to the subspecies *Capra ibex ibex,* but there are three other subspecies recognised: the Siberian ibex *C.i.sibirica* which ranges through northern India, China, Siberia and Afghanistan; the Nubian ibex *C.i.nubiana* of north-east Africa and the Middle East, and the Caucasian ibex *C.i.severtzovi,* restricted to the western ranges of the Caucasus Mountains. The last named animal is often referred to as the Western tur but in appearance, particularly the horns, it resembles more a true ibex than the Eastern or Caucasian tur, *Capra caucasica.*

Of the races of *C.ibex,* those from Asia are the largest, a good Siberian male standing up to 42in at the shoulder with horns of 55in or more. The Nubian ibex, however, although its long slender horns may measure up to 48in is a much smaller animal, a large male seldom being more than 33in at the shoulder.

The body colour of the Nubian ibex is brown with the dorsal stripe, throat, chest and outer surfaces of the legs black, and under parts white. A white band around each leg just above the hooves is a distinctive feature of the subspecies whose present distribution includes Sinai, Palestine, Syria, Arabia, upper Egypt and the Sudan. In Israel it occurs in the Judean desert, extending from the shores of the Dead Sea throughout the Negev Desert, southwards to the shores of the Red Sea at Eilat. In the Sudan it is now confined to the Red Sea Hills.

In Israel the ibex has long been in danger of extinction in the Negev from the firearms of the Bedouin but in recent years, following the passing of conservation laws, they have once again grown plentiful. In former times the ibex was a symbol of beauty and the name Jael, meaning ibex, has been popular from biblical times to the present day. In Arabia the ibex figures in palaeolithic and neolithic rock drawings in the Jebel Tubaiq area and was believed to have been a symbol of the Moon God in the days of the Queen of Sheba; a capitol with ibex horns carved on it was found in the temple of this god at Marib. The Nubian ibex needs to drink quite frequently and is easily ambushed by hunters in the dry season when water is scarce; in some parts of its range, particularly in Arabia, it is in urgent need of protection.

Standing about 38in at the shoulder, the other true ibex is the now rare Wali or Abyssinian ibex, *Capra walie,* frequently classified as another subspecies of *Capra ibex.* It differs from the Nubian ibex by its stouter build, shorter beard and larger, more massive yet slightly shorter horns, and by the possession of a bony protuberance on the forehead. The body colour is chestnut brown, with the chin, throat, under parts and inner surfaces of the legs whitish.

This species is now restricted to the north and northwestern escarpment of Semien, a line of crags which stretches from Adis Gey to Embarass, and is mostly found about the 8,000-9,000ft level, but earlier this century it occupied a larger range and was much more abundant than it is today. The species is threatened by poachers since although it is protected by law in practice this is difficult to enforce. Its habitat is being increasingly invaded by cultivation and domestic grazing animals and although an estimate in 1965 suggested that only about two hundred may still remain, constant vigilance will be needed if the species is to survive.

The markhors *Capra falconeri* of which several subspecies are recognised, differ from other goats and ibexes in the shape of their horns which are straight in direction and appear in two forms—the tight, spiral screw of the Cabul and Suleman races (*C.f.megaceros* and *C.f.jerdoni*) and the open spiral seen in the Astor, Pir Panjal and Chialtan markhors (*C.f.falconeri, C.f.cashmiriensis* and *C.f.chialtanensis*).

The markhor is the largest of the goats; adult males may be over 40in at the shoulder, with long beards extending down the throat and chest to form a heavy fringe or mane, and the horns of Astor and Pir Panjal males may be over 60in in length, measured on the outside curve. The horns of the female are smaller. The long, silky summer coat is reddish brown changing to a greyish shade in the winter, but the fur is not as thick as in the ibex and the markhor is more susceptible to cold.

The markhor ranges over the mountainous districts of Turkestan, Afghanistan, Baluchistan, the Punjab and Kashmir. The Astor and Pir Panjal races (*C.f.falconeri* and *C.f.cashmiriensis*) are found in the eastern part of the range, the former in Baltistan, in the Indus valley, and the latter in Pir Pinjal of Kashmir. In the trans-Indus district of Punjab the Suleman markhor occurs, whilst the Cabul race is found principally in the Kandahar district of Afghanistan, ranging to Baluchistan. In the latter state of Pakistan, in the Chialtan range near Quetta, the Chialtan markhor (*C.f. chialtanensis*) occurs. In many of these localities only a very few individuals remain and protection is desperately needed if they are to survive.

The markhor has decreased in numbers dramatically in

recent years, partly as a result of indiscriminate shooting and also because of competition for limited grazing areas with domestic livestock. Protection, where it is granted, is often ineffective since the animal inhabits areas remote from the observation of game guards even where these exist. In Russian Turkestan two other local races occur, *C.f.heptneri* in the Dashtidjum district of Tadjikistan, and *C.f.ognevi* in the Karluk region of Usbekistan.

The only other species of the genus *Capra* is the unusual sheeplike goat called the tur, or East Caucasian tur, which, as the name suggests, is found in the mountains of the Caucasus. Standing about 38in at the shoulder it is a rather heavily built animal, the general colour being dull brown with the chin, tip of the tail and lower limbs darker, or blackish. The hair is short and there is a small, forward-pointing beard. The horns are almost circular in cross section, but flatter on the back face, and in many respects resemble more those of the Bharal or Blue Sheep than the ibex; indeed the tur is often referred to as the Caucasian bharal. A good pair of horns will measure about 38in along the curve, one of the best recorded attaining 46in.

2

The Origin of Wild Goats

The origins of the various forms of the domestic goat have long been a source of speculation, complicated by paucity of adequate fossil and archaeological evidence and by the difficulty of interpreting what evidence exists. Most authors are agreed, however, in deriving the domestic scimitar horned goats from the wild goat *Capra hircus aegagrus,* variously known in different parts of its range as the Grecian ibex, Bezoar or *Pasang,* which was first domesticated in the Middle East, and which spread over a large part of the Old World during the Neolithic. The horns of these early goats swept straight back over the shoulders in a smooth curve, but goats soon appeared with the horns more or less divergent, spread out laterally, and showing a twist which varies from a slight turn at the tip to a closely wound spiral or corkscrew. Goats having horns showing a moderate twist gradually became commoner and to a great extent replaced the scimitar horned type.

It seems probable that the twisted-horned goats were derived directly from scimitar-horned ancestors by selective breeding, possibly more than once and in more than one locality, but it has also been suggested that the European goats of this kind are descended from an extinct wild goat with twisted horns which is supposed to have existed in south-east Europe in the Pleistocene era. This goat was

22

described in 1914 from fossil fragments and named *Capra prisca* but, according to Zeuner, the concensus of modern opinion seems to be that such a goat never existed.

The markhor possesses horns twisted in the opposite direction to that seen in European domestic goats and it has been shown by breeding experiments that the markhor twist is genetically dominant. It is therefore unlikely that this wild species has contributed to the ancestry of western domestic breeds although it may well be the ancestor of the Circassian goat which occurs from the Caucasus through Turkestan into Central Asia.

The exact date of introduction of the goat into Britain is not known and the archaeological evidence is, in many cases, unsatisfactory as it is only recently that techniques have been developed allowing a clear separation of the bones of sheep and goats on material other than horns and horn cores. In many excavation reports the remains are listed merely as 'sheep or goats'. Nevertheless it is generally accep-

Old English type billy

ted that the goat was brought to Britain sometime during the Neolithic period while Britain was still part of the continental land-mass.

British goats, in the course of time, gradually developed local characteristics and up to about the middle of the last century it was still possible to distinguish English, Welsh and Irish breeds, each having an individual appearance.

David Mackenzie describes the old English goat as a shaggy creature, short on the leg, with horns that swept up, back and outwards in a smooth curve; lactation was long, and the milk had a butter-fat content of about 4 or 5 per cent. Although he suggests that the colour was nondescript the true old English breed was probably a warm blue-grey with a dark stripe running down the back; both sexes had beards, and the male a small tuft of hair on the forehead.

One of the last breeders to attempt to keep the old English strain alive was a lady in Fishbourne, Sussex, Mrs Cartwright, who collected examples mostly from the large commons of Norfolk and Suffolk and from owners who had not had access to goats of foreign or mixed blood. As a billy she obtained a kid from the Cheviots who remained at stud for four years, and by careful breeding a very uniform herd of goats was developed—long-haired, horned, very short in the leg and generally grey in colour. Until 1952 she had had male goats from father to son for seven generations, but during the very wet winter of that year all the goats, except four nannies, died.

The Irish goat seems to have been variable in colour but usually black and white; the most obvious difference between it and the old English breed being in the form of the horns which rose straight and parallel from the brow, turning outwards and a little back at the top in billies, while remaining short and pointed in the nannies. Mackenzie characterises the Irish goat as being leggier than the English, giving a somewhat lower yield of milk with a lower butter-fat content.

The original goat of Wales, according to old descriptions, was. large and mostly white, but by the end of the last century this description hardly applied and Pegler comments that 'it resembles the Irish goat, but is smaller and more symmetrically shaped, the head and horns being lighter and more graceful. The few specimens of the breed to be met with at the present day are not of much value for milk'. The coat of the Welsh goat was said to have more curl in it than the English breed and the horns went straight back instead of sweeping outwards at the ends.

Improved communications initiated the disappearance of these forms and the introduction of foreign breeds which were crossed with British stock contributed to their obliteration, so that it is difficult, if not impossible, to find pure representatives of any of the old local British breeds today.

The foreign breeds which have contributed most to the

lineage of British stock are the Saanen, the Toggenburg and the Nubian; the two hair-producing goats—namely the Angora and the Cashmere—have achieved only park or zoo status in Britain and neither breed has been of any importance in influencing domestic or feral herds.

The Saanen goat takes its name from the Saanen valley in the south of the Canton Berne, and has been widely distributed in Europe, most of those reaching Britain being imported from Holland; single specimens arriving about 1890 to 1900, but it was not until 1922 that the breed was introduced on any considerable scale. The Saanen is shorthaired and ideally pure white, although it is often pale cream or biscuit in colour.

The Toggenburg goat has its home in the districts of Obertoggenburg and Werdenberg in the Canton of St Gall; the colour is light fawn or brown with white markings and the face carries white stripes above the eyes to the muzzle. The coat is short but there may be a fringe of longer hair

Toggenburg male goat

Anglo-Nubian male goat

along the back and down the quarters. Like the Saanen this breed is supposed to be hornless but is only kept so by killing or disbudding those kids that are born with horns. The Toggenburg was probably the first of the Swiss breeds to reach Britain, some half dozen being imported from Paris in 1884.

The first importation of Nubian goats seems to have been in 1883 when a pair were obtained from Paris, having been imported, it was said, from Nubia although Pegler believed that they were more Persian than Nubian. Like the Swiss breeds the Nubian is ideally short-haired and hornless, but it is larger than either the Saanen or Toggenburg, has no uniform colour, and its lop ears and roman nose give it a distinctive appearance.

From these early introductions have developed the British

Saanen, British Toggenburg and Anglo-Nubian breeds, which while remaining basically similar to the original strains, are of mixed origin and in all cases are larger than the pure breeds.

In addition to those mentioned above is a breed developed in Britain and known as the British Alpine. The ancestry of this type is obscure but it is mainly of Swiss extraction and has much in common with the British Toggenburg, being of similar size, long-legged, and having a high milk yield. The coat is short and black with some white markings and, like the other Swiss breeds, is ideally hornless.

While the older feral herds in the British Isles are mostly descended from the original old English, Welsh or Irish breeds, those which have come into being during this century may contain crossbred goats with Swiss blood and some of the old-established herds have assimilated recent additions of Saanen extraction.

In the early years of this century the search for an ideal goat resulted in the production of an enormous variety of crossbred animals. At first these were known as Anglo-

British Alpine male goat

Nubian-Swiss or Anglo-Nubian-Toggenburg etc, names which were self-explanatory but decidedly clumsy, and in 1923 the British Goat Society adopted the term 'British Goat' to cover all pedigree cross-bred goats registered in their Herd Book.

Within historic times there is plenty of evidence of goat-keeping in Britain. In 1229 when Henry III was staying at Stamford he received a petition from men complaining that Hugh de Neville, keeper of Rockingham forest, and his bailiffs had prevented them turning out their goats in the forest of Cliff according to ancient custom, whereupon the king ordered that they should be allowed to pasture them in the more open parts of the forest.

Goats were at all times peculiarly disliked by deer and in consequence were rarely permitted in the royal forests of England, even though other classes of livestock were allowed access to the forest during part of the year. The Scottish law of the forest provided that if goats were found for a third time in a forest the forester was to hang one of them by the horns on a tree; whilst for a fourth offence he was to slay one forthwith and leave its bowels in the place, in token that they were found there.

The tenants of the town of Broughton in Amounderness at the Forest Eyre of 1334-6 claimed to have, from time immemorial, common pasture in the forest of Fulwood (Lancashire) for all kinds of animals save goats throughout the year, except during the mast (acorn season) and the fence month (when the deer were dropping their fawns), by payment of ten shillings (50p) at Michaelmas to the honour of Lancaster. At a swainmote in nearby Wyresdale forest in 1479 eight transgressors appeared for allowing goats into the forest, some of which belonged to the Prioress of Seton and it is recorded that she was fined fourpence for the offence. Similarly in 1323-4 some fifty-six people were presented at the Epping Forest justice seat for keeping goats on the forest contrary to the assize.

Nevertheless, by the seventeenth century goats were apparently prolific in the forest of Kingswood, Gloucestershire, for a survey of 1615 reports, 'Sheepe and goates, most pernitious cattle, intolerable in a forest, make a far greater show than his Majesties' Game.' By this time there were probably feral goats in several forests and Manwood writes, 'There be some wilde beasts . . . that so long as they are remaining within the bounds of the Forest, the hurting of them is punishable by the laws of the Forest, such are wilde Goats, Hares and Connies'.

An early reference to mountain goats occurs in a poem of about 1460 which was sent by Syr Dafydd Trefor of Anglesey to the gentlemen of Snowdonia, asking them to supply him with goats whose milk he required for medicinal purposes. The request did not meet with the approval of Gruffydd ab Tudur af Howell, another Anglesey gentleman, who composed a poem in reply, which, according to the custom of the time was both scathing and humorous and which attacks Syr Dafydd on the grounds that the introduction of goats to Anglesey would be damaging to agriculture and forestry.

Many place names in Britain bear witness to the former presence of goats, especially in Scotland where we find the Gaelic *Garbh* or *Gabhar,* meaning a goat, in combinations such as Stob Ghabhar (hill of goats) on Blackmount and Uamh nan Gabhar (goat cave) on Colonsay, while Coire nam Meann (corrie of the kids) is not uncommon. Feral goats were doubtless common in Scotland from the earliest times and in Perthshire, during the rebellion of Huw Murray (1661-79) it was said that he and his men would never be subdued so long as they could get a wild goat at Creag Mhor.

The goat has been called 'the poor man's cow', and

certainly no better description could have been applied in the days before the clearances. In the mid-eighteenth century it is estimated that the human population of the Highlands was about ten times what it is today; the standard of living was very low, the peasants relying mainly on goats for meat and milk. As an example of the stock that might be held by a farmer of this period we may note the possessions of Rob Roy, listed in his will of 1738, which shows 12 cattle, 12 sheep, 23 goats and 5 horses. During the summer the goats, along with any sheep or cattle, were driven away from the arable land about the crofts and herded from summer sheilings in the mountains by children who kept a watchful eye on them, in much the same way as peasant children still do today in parts of Europe. At night the goats were driven into bothies and bedded with bracken or moss, which by the spring contained a large accumulation of manure, said to be excellent for assisting the growth of potatoes.

As Mackenzie points out, the tendency of the goat to go feral evinces itself most at the onset of the rut and at kidding time and thus, in the Highlands, the tendency to stray occurred at seed time and harvest when there was a minimum of labour available to prevent it.

At one time it was the custom for large herds of goats to be driven from the Lowlands of Scotland northwards, the owner selling his animals as he went: if business was brisk he might have sold all his animals by the time he reached mid-Scotland, but if not he might end up in Caithness. When travelling such long distances straying would be difficult to prevent—the wild goats of Slochd (Inverness-shire) are thought to have originated in this way.

The drinking of goat's whey was at one time as popular as 'taking the waters' at health spas and, as Irish goats were famed for their milk, importations were frequent. One of these whey-drinking centres was established at Leswalt in Wigtownshire, but by the beginning of the last century, it had been disbanded, as had the goat-milk spa at Blairlogie, near Stirling. Other whey-drinking spas were situated in the Clackmannanshire foothills and on the Isle of Arran, while in Wales, Abergavenny was a popular resort in the eighteenth century.

To reach the whey-drinking centres new intakes of goats would often have to travel long distances, especially if they had been shipped from Ireland, and as large droves were driven through Wales en route for England doubtless some escaped to wander off into the mountains. One of the last droves to pass through South Wales consisted of some 300 beasts which were driven from Cardigan to Kent in the autumn of 1891; it took three men, three boys and five dogs to control them, but how many escaped on the way is not recorded.

Mackenzie writes:

Irish goats were annually imported and distributed through the hill districts of Britain in nomadic droves from which the milkers were sold as they kidded. Up till the 1914 war the Irish goatherd, shouting picturesque advertisement of his wares, squirting great jets of milk from his freshened nannies up the main street, was a regular harbinger of spring in the mountain villages.

The importation of goats from Ireland still continued after World War I, and as late as 1926 the total of goats exported from Leinster, Munster, Connaught and Ulster into England was 241,427 head, a decrease of less than 10 per cent on the figures for 1881.

Improved agricultural methods reached the Highlands about the middle of the eighteenth century and as the breeding of the improved Border sheep on the hills became popular so did the interest in goats wane. The discontinuation of whey drinking was another, though minor, contribution to the decline, and when this custom finally died out many of the big herds of goats may have taken to the wilds where this was possible. The migration of people to the new industries of the towns left many of the isolated crofts empty, and the goats which formerly bedded in the bothies ran wild in the mountains.

Paradoxically the slump in sheep farming at the end of the last century did nothing to help the goats; the hills were cleared of sheep to give the stalker the best chance of coming to terms with the stag and if sheep were a nuisance to stalking many thought goats worse and they were consequently often shot.

David Mackenzie relates the tale of a crofter who kept a billy with his cows to frighten off the deer, the idea being that the stink of the goat would mask from the deer more delicate scents such as the approaching stalker and so undermine their sense of security that they would keep clear of the whole area; if the fable was popular, he remarks, it must have killed some goats.

Not every forest owner, however, considered that the only good goat was a dead one and many feral herds were deliberately preserved, fresh blood sometimes being introduced to prevent the stock becoming too inbred. There are also many cases known where goats have been deliberately introduced, sometimes to provide sport, or simply to add to the interest of an estate. The goats at Lynton Rocks in Devon, for example, were introduced by the original owner to enhance the rugged beauty of the valley.

The origins of the wild goats of Britain are thus quite prosaic but more romantic explanations, ranging from the plausible to the incredible, also exist to account for the origins of some herds. The goats of the Cheviot Hills, according to local legend, are derived from stock turned

White goats on the Island of Cara in 1939

loose on the mainland by monks from the priory of Lindis-
farne and the belief that many of the herds on the west
coast of Scotland, including the white goats of the island of
Cara, are descended from goats carried by wrecked galleons
of the Spanish Armada is widespread.

It does seem probable that goats were carried on the ships
of the Armada as a source of meat and milk but the goats
of Cara are typically Saanen and bear little resemblance to
any present day Spanish breed. Mackenzie, however, points
out that at this time the Netherlands, where the white goat
of Swiss ancestry has a long history, were part of the Spanish
empire and it is therefore possible that the galleons carried
goats of this type. The same author makes another suggestion
concerning the origin of the white goats of the Scottish coast:

It is equally likely that they owe their origin to Scan-
dinavian seafarers who frequented these shores, in whose

homeland the white 'Telemark' goat has long been popular. The sea route to the Western Isles was assuredly more hospitable than the land route until the late eighteenth century. Many of these feral flocks exist on small islands. But goats are bad swimmers; the goats of Ulva were exterminated by being driven out on to a tidal reef when the tide was rising. The ability to swim a hundred yards would have saved them. So it is highly improbable that these island goats swam to shore from wrecked ships. The prevalence of the tradition that they did so suggests that their origin is wrapped in mystery and antiquity. If their existence were due to the obviously sensible practice of sending dry stock and males to uninhabited islands for the summer—to save herding them from the crops—then the mystery would not exist and the colour of the goats would not be so prevalently white. It is tempting and not unreasonable to suggest that the Viking longboats which carried cattle to Greenland in AD 1000 and pirated West Highland waters for centuries, may have carried the white goat of Norway to the Western Isles, and their islet strongholds. It is altogether appropriate to believe that the Vikings sustained their heroics on a diet of goats' milk and kid. In any case, there can be little doubt that the native goat stock along the whole seaboard of Britain was liberally mixed with 'ship-goats' from abroad.

3
Natural History of the Wild Goat

One of the first descriptions we have of the wild goat can be found in Turbervile's *The Noble Arte of Venerie or Hunting* (1576):

The Wilde goate is as bigge as an Harte, but he is not so long, nor so long legged, but they have as much fleshe as the Harte hath . . . They have a great long beard, and are brownish grey of colour like unto a Wolf, and very shaggie, having a blacke list all alongst the chyne of their backe, and downe to theyr bellie is fallow, their legges blacke, and their tayle fallowe.

Owing to their mixed origins there is no fixed type of colour or pattern in the wild goats of Britain and they may vary from jet black to pure white, although most are a mixture of brown, grey, white or black. Some of the darker animals have small tufts of white hair on the forehead, whilst others may have a black or grey forequarter and a white rear quarter; silver-grey, brown, dun or yellow are among other colour varieties and the nannies appear to be more prone to colour variation than the males. Most truly feral goats, especially the billies, develop long shaggy coats, but one occasionally comes across a short-haired animal running wild.

Fraser Darling points out the potency of natural selection in levelling the type of goats newly gone feral and suggests that the reversion to wild type is rapid, ten years making a

Wild billy on the Grampians in 1950

big difference to the general appearance of a herd; he also
wonders what the influence can be on nitrogen metabolism
that makes feral goats run increasingly to hair and horns
until the standard of the wild goat is reached.

Goats, like sheep, are social animals tending to live in
herds whose size and composition varies according to the
species and the time of year. In British feral goats the old
billies keep together in groups of three or four through most
of the year, keeping quite separate from the nannies except

during the autumn. The nannies and young animals form a separate herd which usually sticks closely to a definite territory whereas lone billies, especially during the breeding season, may wander as much as 50 miles from their home.

Goats rise to their feet like deer, raising themselves first on their knees before standing up; when lying down the process is reversed.

Some breeds of goats have tassels under the jaw, the exact function of which is not known. I have never seen these appendages present on truly feral goats.

It is seldom that goats panic like sheep, and rush off when seeing a human being approach at a distance; without appearing to be in any particular hurry their retreat is made in an orderly fashion, and in a surprisingly short space of time they will have disappeared from view. When retreating, particularly if alarmed, goats will often raise their tails like deer in similar circumstances.

The social structure of natural goat herds has been studied by several workers, most of the available information being summarised by Hafez. Within the herd each individual has a place in the hierarchy and 'knows his place' being dominated by animals higher in the order than himself and in turn dominating those lower in the scale. In dominance relationships the most important factor is age; kids approaching adults other than their mothers are butted away and quickly learn to be subordinate—hence old animals are, in general, dominant over young ones.

In unrelated goats of similar age dominance depends on relative size and strength, the possession of horns conferring an advantage, and it is found that males, being of superior size and having larger horns, are usually dominant over females, although a hornless male is seldom a match for a horned female. The most dominant animal is not necessarily the leader of the herd since leadership is based on different criteria, the most important being the relationship between mother and young, and the oldest nannies with the greatest number of descendants tend to become leaders even though the oldest and strongest males are usually most dominant.

The dominance relationships in the herd greatly reduce the amount of actual fighting that occurs between individuals since once the relationship is established it is long-lasting and mere threat gestures are usually sufficient to enforce the rights of the individual. Fighting is not, however, entirely eliminated and may often take place between well-matched males, especially during the rut, the animals rising on their hind legs, throwing back their heads and hurling themselves forward to meet with a loud clash of horns.

It was Hingston's opinion that this behaviour, in contrast to that of sheep, which rush at each other and meet head-on without rising on their hind legs, was connected with the possession of well-developed beards which the billies use as organs of threat—the animal having the best beard intimidating his less well-endowed rivals.

Whatever its purpose the beard of the male goat undoubtedly enhances its appeal, and poets have not been slow to note the appendage; Tennyson refers to the 'beard-blown goat' whilst an unknown poet gives this delightful description of a bearded billy resting on some windswept crag:

Hung high in air, the hoary goat reclined,
His streaming beard the sport of every wind.

A writer of the sixteenth century suggested that a simple method to prevent a goat from straying was to cut off its beard!

The rutting season occurs in the autumn from about the beginning of September to the end of October, the most important factor in bringing the onset of the rut being diminishing daylength. Hence Mackenzie has produced a map showing that in the north of Scotland the breeding season starts about the 10th of August; about the 2nd of September in the Cheviots; the 16th of September in mid-Wales and during the latter part of September in south Cornwall.

The goat will, of course, breed at all seasons of the year, and whilst the dates given by Mackenzie may mark the beginning of noticeable activity among billies, the main rut seems to take place about a fortnight later. The rut lasts for about a month but the billies do not collect the nannies together as a red deer stag collects hinds, but behave more like a ram with the ewes. Mixed herds form, of twenty to thirty animals or more, made up of all ages and sexes and including often two or even three big billies. The big billies do not seem to fight very much, though there may be a certain amount of infighting; that is to say, getting close and

Billy goat approaching a nanny during the rut

then pumelling each other in the ribs, but not actually backing apart and meeting head on. One has the impression that the billies do not bother each other very much, and in all probability they will often serve the same nanny. Occasionally one sees a head-on fight or an old male chasing off a younger one, but on the whole the billy does not guard his wives as jealously as the stag.

When a billy approaches a nanny in season he stands alongside her, often curling a horn over her and rubbing her back with it; occasionally he makes a rumbling noise which one has to be fairly close to hear, the billy pushing his nose against the neck of the nanny, close to her ear, while making it. Generally speaking, however, the billies are not particularly vocal during the rut but at any time of year wild goats will give an explosive snort through their nostrils when alarmed, and on a still day this sound may disturb deer resting in a corrie perhaps half a mile away. This sneezing-like sound is also used by the billies during courtship, and whilst the noise is being emitted the whole back of the body gives a quiver.

Sometimes a billy will give a grunt of annoyance. On one occasion I saw a nanny escape from the attentions of a billy by getting under the cover of some rocks; the old billy frantically tried to scratch her out, making this peculiar grunting noise the while. Although it is comparatively rare to see a stag covering the hind I have several times seen a wild billy mating a nanny.

In no domestic animal does the instinct of reproduction develop so early as it does in the goat, Pegler stating that he had known instances of kids being mated at three months old whilst still being suckled by their dams, and producing a live kid in due course: billies will also breed when under a year old.

The gestation period of the goat lasts about 150 days, the

Kid among rocks, Craig-y-Benglog, Merionethshire in April 1952

bulk of the kids being born between the end of January and the end of March—generally a most inhospitable time of year for any young animal to be introduced to the British climate. During the first day or two after birth the kid will generally be left by its mother, who will only visit it at feeding times; she never wanders very far from the kid, however, until such time as it begins to follow her.

Twins are not uncommon but I have never seen triplets in a feral state. When the birth of kids coincides with a severe wintry spell, mortality is high and it is this factor more than any other which is responsible for the steady population level maintained by herds in the more northerly parts of Britain. Severe snowstorms can also play havoc with adult wild goats and dozens of skeletons have been found in the sheltered parts of the rocks on Glentarroch where the goats have taken refuge during a storm and been smothered. It is well known that goats dislike the wet and it was Mackenzie's opinion that in the Scottish highlands 'the feral goat population is very much limited to the number of dry beds available on a wet night'.

Kids may often get trapped among large boulders or on rocky screes. On Rhum, for instance, one often finds young kids lying dead on their tummies, with their small legs hanging in the crevices; in coastal localities where the goats often walk out on to the shore to eat seaweed, the kids frequently follow their mothers, doze off on the rocks, and are cut off by the incoming tide.

Foxes also take a number of kids, though if the nanny is at hand she will generally try to defend her youngster. In the Cheviot Hills the College Valley Foxhounds, during the

Twin kids on Craig-y-Benglog, Merionethshire in April 1952

spring hunting, occasionally pick up and destroy a young kid, but otherwise the hounds are steady on the old goats as it has been the custom of the Master to keep a tame billy in kennels to stop hounds from riot. He maintained that wild goats have a gamey odour which is tempting to foxhounds unless they are used to it.

In Scotland quite a number of kids are killed by the golden eagle, as a visit to any eyrie in a goat locality will confirm; evidence is not confined to the eyrie for remains such as legs or the forehead of the skull may be found miles away to show where an eagle has had a good feast. Duncan McNaughton, the former head stalker on Rhum, once told me that he believed the eagles will, at times, be responsible for putting full-grown goats over the cliffs, though he had never actually witnessed such an event.

The horn, of course, is the principal feature of most wild billy goats, and but for this trophy few people would ever bother to stalk him for sport. There are two main types of horn formation; the first—which I think the most attractive—is that in which the horns go backwards for perhaps 12-15in and then spread out sideways rather like a Spanish ibex. A good head of this kind will have a horn length of about

The age of the goat shown by the horn

30in and a similar spread; the beam at the base should be about 8in in circumference. In the other type the horns curve straight backwards in similar fashion to the Grecian ibex, and here, a length of 30in would be a good head. The best Scottish head, from the Isle of Bute, has a length of 44¾in.

Examination of the horn will enable an accurate estimate of the age of the goat to be made, for a new segment is grown each year and by counting the segments the age is known (see page 44); the segments are separated by rings which represent the check in horn growth in the winter and this method of ageing is reliable in some of the other bovids, including the European ibex.

The following diagram, showing the development of the incisors in the goat was prepared by the late H. Stainton.

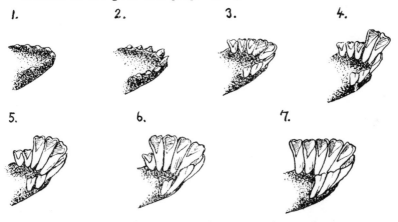

1.   *2.*   *3.*   *4.*

5.   *6.*   *7.*

The age of the goat shown by the teeth:
1, *at birth;* 2, *at* 36 *days;* 3, *at* 12 *months;* 4, *at* 14 *months;* 5, *at* 2½ *years;* 6, *at* 3½ *years;* 7, *at* 4½ *to* 5 *years when they are full mouthed*

The dentition may also provide a guide to the age of a goat since the incisor teeth appear in a regular succession over a period of years. The permanent incisors are usually

45

cut between 12 and 14 months and by the time the goat is $2\frac{1}{2}$ years old it should have four large incisors; at $3\frac{1}{2}$ years of age there will be six incisors whilst between $4\frac{1}{2}$ and 5 years old it becomes 'full mouthed' having eight large incisors in the lower jaw and three premolars and three molars on each side, top and bottom, making a total of thirty-two teeth. This may be compared to the red deer which has its full complement of thirty-four teeth at $2\frac{1}{2}$ years and the roe which is full-mouthed at about 14 months.

In a domestic state, goats—particularly the nannies—have often exceeded 16 years of age but I believe feral billies seldom exceed 12 or 13 and have never seen a truly feral billy in excess of this age.

The average weight of a good Scottish wild billy will be between 8 and 10 stone, though weights will of course vary somewhat from district to district; two 7 year old billies, shot on Cairnsmore in Kirkcudbrightshire, weighed 7 stone 12 lb (110 lb) and 7 stone (98 lb) as they fell. Very much greater weights are claimed; Mr W. Joynson once told me that he shot a billy in Corrie Na Urisgean on Ben Venue, Perthshire, in 1937 that he estimated at 22 stone (308 lb) and Kenneth Richmond suggests that some of the Ben Lomond goats may weigh up to 25 stone. There are undoubtedly some larger than average goats in the Ben Lomond district but I very much doubt that any reach 20 stone; the best goat reported by George Jones, keeper at Rowardennan on the south-west face of Ben Lomond, weighed 12 stone 5 lb, and it was his opinion that there would be few over 12 stone.

The late David Mackenzie, replying to my enquiry, wrote that in his opinion feral goats of Galloway weigh about 80-85 lb for females and 100-120 lb for males; Carradale goats slightly heavier, the males up to 170 lb. He mentions a mongrel domesticated male, genetically similar to many feral types, encountered near Burton-on-Trent which weighed 19 stone 12 lb, the biggest goat he had ever seen and as thick as a fat shetland pony.

The same authority also points out that normal domestic billies reach a top weight of 17 stone 12 lb, and exceptionally up to 21 stone 6 lb, such animals lacking horns, being short-haired and short-lived; he goes on to say:

> I think, therefore, that the 23 stone hill billy is a possibility; 20 stone I could easily believe. He would have to be a late kid. I think it is rash to judge by the Welsh coast types, which are very short of minerals, or by the northern types which breed very early with resultant malnutrition in the first few months of life. Furthermore, it must be remembered that goat weights of all kinds must be approximations owing to the huge capacity of the digestive organs. A thirsty goat of 100 lbs can put on 25-30 lbs (say 2 stone) in five minutes at a water point.

Despite these views I would be extremely surprised if any truly feral horned goat exists in Scotland whose weight exceeds about 12 stone, the majority being in the 7-9 stone class. An average feral goat with a shoulder height of about 32 in is a considerably smaller animal than a highland red deer stag whose height is about 43 in and whose *average* weight is not much more than 14 stone gralloched. A 20-stone stag, let alone one of 23 stones, is indeed a prodigious animal. If 20-stone goats exist on the Lomond hills this would surely indicate very rich feeding and the deer would likewise benefit; but this is not the case, for the deer weigh about the average.

Although goats and sheep are frequently found on the same mountain, crossing between the two, although possible, is rare. Some years ago it was reported that fertile hybrids were being produced in Bulgaria by Professor Bratanov who hoped to produce an animal having the milk yield of the goat and the wool clip of the sheep. A 'shoat' seems to be an appropriate name for such a hybrid! Goat/sheep hybrids

47

Billy goat—a cross between a Saanen and a wild Scottish goat

have also been reported from Malta and (by C. H. Shaw) from Puffin Island, off Angelsey. Warwick and Berry, however, in 1949 reported that such cross-species matings were always sterile, the embryos dying early in development. Certainly, when the two animals have access to their own kind cross-matings are rare or non-existent but if the billy or ram is denied access to the appropriate female then it is a different matter. The fable that goats will interbreed with the sheep and ruin the stock is, however, still prevalent among the sheep fraternity and the goats of Ulva, off the west coast of Argyllshire, were slaughtered under this pretext some years ago.

In Europe crosses between feral domestic goats and wild ibex have often been recorded, the crosses being themselves fertile. In 1947 during a chamois hunting trip to Austria, I met an old Jäeger called Fritz Thaler who had in his house a number of goat skulls each carrying four horns which, he informed me, were the result of a chamois cross. His story was as follows.

A number of years previously he had witnessed a wild chamois buck join a small herd of tame goats and mate with one of the nannies. The resultant offspring was a billy with four horns, who in due course mated with the nannies and produced two nannies and a billy, all with four horns. From what I could gather the trait was carried on through the male rather than the female line and up to 1947 four-horned animals still occurred in the herd, though unfortunately lack of time prevented my seeing one in the flesh. Needless to say these animals caused considerable local interest and one of the billies soon found himself a member of a travelling circus.

In an attempt to throw further light on the possibility of a chamois/goat cross, Mr T. C. S. Morrison-Scott at the British Museum of Natural History drew my attention to Dr Couturier's fine monograph on the chamois in which he records one which, during the rutting season, followed a herd of goats into a little village in the Tarentais and was captured by the villagers. The same author stated that it was a fairly common occurrence for a domestic goat to run wild with a herd of chamois and then become very wild itself.

Charles Boner on the other hand, in *Chamois Hunting in the Mountains of Bavaria* (1853), concludes:

But the most decisive proof of the non-affinity of the two animals is that they never generate together. Although in the mountains herds of goats are constantly wandering about near the haunts of chamois, no one instance is known of a she-goat having brought forth young which were a cross between the two breeds. The chamois indeed always avoids the places where goats have strayed.

If these observations are reliable then the possibility of a goat/chamois hybrid would appear remote; from my own limited experience I can only say that I have several times seen chamois and domestic goats within 200 yd of each other, but not mixed.

A skull I brought home with me certainly had no visible chamois characteristics; the frontal bone is convex while that of a chamois is concave; the four horns all recede, the foremost pair being flat at their base (a typical feature of goats) while the horns of the chamois are round and always advance. Mr Morrison-Scott, confirming this, remarked that the horns follow the scheme for four-horned sheep, the frontal pair being normal and the back pair under-developed and curving downwards and backwards.

E. R. Alston recorded that in 1879 a chamois skull with four horns was exhibited at a Zoological Society meeting by Rowland Ward who observed that although the specimen had been injured and carefully repaired the frontal sinuses and bases of the horn cores were uninjured, and that in his opinion the specimen was genuine.

Apropos of multi-horned chamois, W. A. Baillie-Grohman wrote:

> In continental collections one now and again is shown chamois heads adorned not with two, but with four or six horns. If the owner is not aware of the prank that has been practised on him, he will declare it to be a rare and valuable monstrosity. But it is nothing of the kind, only a cleverly contrived swindle by which many a collector has been caught. The several chamois horns are genuine enough, but the skull is a perfectly bona fide frontal bone of a sheep—the Sardinian species which had generally four horns and often six horns.

Another common trick was to mount a small pair of chamois or roebuck horns on a hare's head with such a high degree of skill that many travellers to the Tyrol and elsewhere, on seeing such a specimen decorating an inn wall, immediately jumped to the conclusion that horned hares existed in those high altitudes!

Paul Dalimier in his paper on the morphology of the domestic goat remarks, however, that supplementary horns are a fairly common anomaly among goats and animals with three or four horns are not rare, the record number being eight.

The goat is a browser, not a grazer, preferring to browse on the leaves and twigs of shrubs and coarse weeds rather than lush pasture. It is probably the most destructive of all domestic animals and it has been said that the passing of a herd of goats is as ruinous to young trees and shrubs as a

Young goat browsing

bush fire. In goats, as in ibex, the two toes can be separated and then pressed strongly together to provide a grip on the smallest projections enabling them to traverse almost vertical rock faces offering the minimum foothold. This faculty also enables goats to climb trees and in parts of North Africa they live a largely arboreal existence in the flat-topped acacia scrub some 10-20 ft above the ground. In Morocco and in particular around Mogador, Agadir and Tarondant I have seen the native goats climb to the upper branches of the *arganier* trees; habitat necessitates these acrobatics for the ground is so arid and stony that there is to all intents and purposes no green except the arganiers and some rather unpalatable camel bushes.

A herd of goats will clear land of bramble, briar, ivy, gorse and heather — in fact pretty well everything of a scrubby or woody character and are fond of thistles, docks, stinging nettles and other intractable weeds. They were doubtless of great use to early pastoralists in clearing forest and scrub which was then available for grazing animals or growing crops.

During the 1930s the number of goats in the Burren hills of Ireland probably ran into hundreds but during and after

Goats eating gorse

the war great numbers were rounded up and exported for food; kids at this time were fetching about two shillings apiece. It was not long before the local people began to regret the shortage of goats, for the scrub soon started to spread and became well-nigh impenetrable in parts where their cattle used to graze.

When the island of St Helena was discovered by the Portuguese in 1502 it was covered with a luxuriant forest. Goats, introduced in 1513 increased rapidly and the native plants and forest began to disappear; nevertheless the native ebony continued to be abundant and some 200 years after the introduction of goats was still being used as fuel for burning lime. By 1810, however, when the older trees had all been felled, it was realised that new forest growth had long since ceased and the island had become a barren and rocky waste.

The same story has been repeated many times the world over and it is thought that overgrazing, particularly by goats, has played a major role in the creation and extension of many deserts, including the Sahara. The establishment of feral goats has had more or less disastrous effects on the ecology of Cyprus, New Zealand, Hawaii, several of the Caribbean islands, and on Guadalupe island in the Pacific, destroying native vegetation, causing soil erosion and irreversible damage including the extinction of much flora and fauna. A few years ago damage to cultivation by goats on Lebanon was estimated at about six million pounds — a heavy bill to pay compared to the mere three-quarters of a million pounds the keeping of these animals brought to their owners.

Where wild goats have access to woodland in Britain they do a considerable amount of barking, particularly during the winter when there is snow on the ground. Their favourite trees seem to be holly, ash, elm (especially young

Trees barked by goats in Bagot's Wood

trees), rowan, hazel, willow and yew. It is probable that oak
trees will also be attacked but it was noticeable that the
Bagot goats (page 99) who had a wide choice appeared to
miss out the birch, oak and older elms. When timber is being
felled, or when trees are blown down in gales, goats will
eat the buds and twigs of almost any tree they find.

The bark of the yew does not seem to harm either goat or
cattle, for cattle have stripped the bark off most of the yew
trees along the east side of Windermere, apparently without
ill effect. Colquhoun wrote of goats on Crapna-gower 'busily
engaged in cropping the leaves of a venerable yew munching
their delicious repast of yew twigs'. He also described a fine
old goat who hanged himself by the horns on a yew tree, in
attempting to feed on the higher branches.

Although the goat, like the roe deer, can tackle the living

yew twigs if it eats dead yew branches it will probably suffer violent sickness: unripe acorns, hemlock roots in winter, rhododendron and laburnum are also said to be poisonous to goats and three Bagot goats taken to Cwm in 1954 all died the following spring, believed poisoned by rhododendron. The wild goats around Dochfour are fully aware of the poisonous character of this shrub and although they come down regularly in winter into woods which are full of rhododendron they never touch it.

Goats appear to be fond of salt and in any area which has access to the seashore they will often be seen climbing about the rocks and eating the seaweed; this habit has been the death of many goats, for being indifferent swimmers they have at times been cut off by an incoming tide and drowned.

Wild goats seem to carry very heavy burdens of parasitic worms but apparently without much ill effect; some thirty-eight beasts (all but two from Scotland) examined by Dr A. M. Dunn of the Glasgow University Department of Veterinary Pathology, yielded nineteen species of parasites, mostly from the abomasum and small intestine. Dr Dunn reported as follows:

A very common parasite was *Haemonchus contortus,* one of the abomasal worms. The general importance of this worm is that it is a bloodsucker, capable of causing very severe anaemia and often death. The two curious features of its presence in these feral goats are, first, the tolerance by the animals of very heavy infections and, second, the large numbers present when this parasite is quite rare in British, and especially Scottish, domesticated stock. It is difficult to propose any explanation for the heavy burdens in the feral animals, but one may suppose that, since this worm is very susceptible to anthelminthic drugs, it has been kept to low numbers in domesticated stock by periodic treatment; the feral animals are not husbanded and the infections may grow unchecked to very large proportions.

55

Three of the abomasal species had never previously been found in goats. These were *Ostertagia leptospicularis, Grosspiculagia lasensis* and *Spiculopteragia spiculoptera*. They have only been found in goats which share grazing with red deer and, in fact, deer are the prime hosts of these three parasites. Though their occurrence in goats is of somewhat academic interest it does demonstrate that they are capable of passage in animals other than deer, and hence they may be expected eventually to be found in any ruminant which shares grazing with deer; the Scottish hill sheep, which often share wintering with deer is an obvious subject. The dominant species of parasites in the feral goat in this country are *Ostertagia circumcincta,* which is also a common parasite of domesticated sheep and goats here, and *Trichostrongylus capricola* which in this country is exclusively a goat parasite, though it is common enough in sheep in countries where they graze communally with goats.

The parasite populations in these feral goats which, by the way, were all killed by shooting and were not culled on a basis of poor condition, have approximately the same species composition as those of domesticated sheep, and there is a possibility that where sheep and feral goats occur together, as in some of the island herds, the goats may act as reservoirs, and their high parasite burdens complicate the control of parasites in the sheep stock.

Goats are very often infected with cysts of *Cysticercus tenuicollis* the intermediate stage of a dog tapeworm; the cysts closely resemble the gall bladder, which is present in the goat but absent in red deer, and in consequence deer stalkers, who are familiar with deer lacking a gall bladder, have on occasions treated this organ as an abnormality in other animals.

The ears of feral goats may be affected by skin lesions and in goats from Bagot's Park, Staffordshire and Dumfriesshire the causative agent was identified by Dr Christine O. Dawson as a fungus, *Peyronellaea glomerata*. This fungus is apparently ubiquitous on all kinds of vegetation from which susceptible animals pick it up—there being no question of goat-to-goat transmission. Two factors appear to aid infection of the ear, firstly the situation of the ear which is prone to minor trauma and the fact that it is less densely covered by hair, and secondly the surface temperature of the ear, which is lower than most other parts of the body thus allowing the proliferation of the fungus to which a temperature of 37 C is inhibitory.

Ringworm is comparatively uncommon in goats, *Trichophyton verrucosum* appears to be the most frequent species and an epidemic in domestic stock has been described in which 83 per cent of the kids and 3.5 per cent of adults were infected, lesions being confined to the head.

Among the ruminants the goat is particularly odorous, the smell being mostly produced by glands situated immediately behind the bases of the horns. These are present in both sexes but are only activated by the presence of male hormones in the blood stream; hence it is normally only the billy that stinks and the smell is greatly accentuated during the rut, when the billy often intensifies his natural odour by micturating upon his forehead, bending his head between the forelegs to receive the urinary spray. I have frequently seen a rutting billy do this and Pycraft states that captive goats may carry the habit to such excess that blindness results.

Stephen Holmes notes that the smell 'in an old goat may be distinguished half a mile distant, and a person touching the creature with his hands or his clothes will find the smell hang about him all day'. Even when dead the smell persists and it is seldom possible to hang a mounted head in a living room, as many sportsmen have discovered.

4

Goat Lore and Legend

Goats have long been associated with many curious practices and ceremonies in religion, witchcraft and folklore and much of the early lore is entertainingly summarised by Topsell, whose celebrated seventeenth-century compilation *The History of Four-footed Beasts* devotes twenty-four pages to this animal.

According to the old Jewish ritual, on the great day of atonement, the sins of the people were laid on the head of a goat which was sent into the wilderness as a 'scapegoat'. The scriptural he-goat, as given by one writer, was 'the sin-offering for sins unwittingly committed'; and for sins of the congregation on the day of atonement when one goat was sacrificed and another (Azazel), dyed with its fellow's blood was driven over a precipice as a symbol of pardoned sin.

Some interesting information is provided by D. A. Mackenzie, who writes that the goat was also associated with Merodach, the mythical Bel or 'Lord' of the Babylonian Pantheon. 'Babylonians, having prayed to that god to take away their diseases or their sins, released a goat which was driven into the desert . . . In India the goat was connected with Agni and Varuna: it was slain at funeral ceremonies to inform the gods that a soul was about to enter heaven.' Apparently Babylonian astrologers believed in a celestial mountain with Polaris situated at the summit—Polaris being identified with the sacred goat 'highest of the flock of night'. Tammuz, like Anshar, as sentinel of the night heaven, was a goat, as was also Nin-Girsu of Lagash.

Indeed the astral honours of the goat are rather exceptional. The Milky Way, 'the bridge of souls', is sometimes referred to as the she-goat, whilst the Pleiades are called, in many parts of Europe, the Seven She-goats.

The fabulous monster called 'Uraisg' or 'Uruisg' was half goat, half man—a satyr in short. The satyrs, mythological attendants of Dionysus, were forest gods associated with fertility rites, having the hindquarters of a goat, small horns and goat-like ears. From the earliest times the goat has also been associated with the devil who not infrequently appears in the form of this beast, and there is an old country superstition that no he-goat ever remains in sight for twenty-four consecutive hours because once a day it pays a visit to the devil to have its beard combed.

The origin of the oft-repeated phrase 'separating the sheep from the goats' is of course biblical; St Matthew writes of the final separation of the righteous and the wicked on the day of retribution. However, when given the chance to redeem its bad name it seems the goat scorned to do so, or so we learn from a rather sad Finnish legend about Christmas. In Finland, not so long ago, children had their presents brought by a goat dressed in a thick, shaggy fur coat instead of the conventional gentleman dressed in red who comes down the chimney. The legend tells how a goat was in the stable when Christ was born. It was bitterly cold and Mary asked the goat to spare some fur to make a shawl for the baby, but the goat declined, whereupon an angel told him that because of his selfishness he would be compelled to walk about the world in the snow every Christmas night carrying gifts for little children. The only way in which he could be released was for a child to stroke him and thank him for his gifts; but according to the legend no child ever did and the goat therefore had to spend every Christmas night atoning for his meanness.

There seems to be no ready explanation of the phrase 'getting one's goat'—an old Americanism for annoying someone, though in 1959 there was some correspondence in *The Field* about it. One writer suggested that it arose 'through the practice of an American golf club issuing members with a medal inscribed with a goat which the loser gave to the winner after each round . . .' Perhaps another correspondent was nearer the truth when he asked 'why, in the deep past, should not it have been said in England of an indignant non-commoner who found his goat in the village pound for trespass on the village common? . . . down at the pound, they got oad Tom's goat agen'.

The 'giddy goat' seems a singularly inappropriate phrase as it must be obvious to anyone who has watched goats in the wild that they can never experience the feeling of giddiness—or few would survive. It may of course refer to their antics in steep places, which make the onlooker giddy to watch.

Porta gives an unfailing recipe for catching the sargon (a sea-bream much esteemed by the Romans) by taking advantage of its affection for the goat. 'Sargi love goats unmeasurably . . . when so much as the shadow of a goat that feeds neer the shore shall appear neer unto them they presently leap for joy and swim to it in haste . . . the Fisher, putting on a Goat's skin with the horns, lies in wait for them, having the Sunne behind his back and paste made wet with the decoction of Goat's flesh: this he casts into the Sea where the Sargi are to come: and they, as if they were charmed run to it, and are much delighted with the sight of the Goat's skin and feed on the paste. Thus the Fisherman catcheth abundance of them.' Another association of the goat with fishermen was a custom of the islanders of the western islands of Scotland up to about the seventeenth century who would hang a he-goat to the mast to procure a favourable wind. (I cannot believe, however, that it was any such ancient custom that prompted students of Merton College, Oxford, to tether, one June evening in 1958, a billy goat to the balustrade around the flat roof of the college tower. The beast was eventually rescued by an RSPCA inspector, who had the unenviable task of bringing it down some 167 steps of a spiral staircase.)

The goat figures commonly in fables, including many of Aesop's where he is sometimes represented as being none too clever; Reynard the fox, on being assisted by the goat out of a well, remarks 'If you had half as much brains as you have beard, you would have looked before you leaped.' Whatever Aesop's opinion, there is no doubt that the goat is an intelligent creature and several early writers comment on this, including Pliny who relates an incident reported to him by Mutianus who witnessed the event: two goats met on a narrow bridge which would not allow them to turn round or go safely backwards, so one of the animals lay down flat, while the other walked over it.

There are a number of fables concerning the goat and the wolf which exist in several versions. One variation of a Russian story relates that a goat is about to give birth under an apple tree which advises her to go elsewhere lest the apples fall on the newborn kids and kill them. The walnut tree gives similar warning, upon which the goat occupies a deserted tent in the forest and brings forth her young. When the goat goes out she cautions the kids not to open to anyone, and when the wolf comes they know from the roughness of the voice that it is not their mother and refuse to admit him. The wolf then goes to the blacksmith and has a voice made for him resembling that of the goat; the deceived kids open, and the wolf devours them all except one who hides under a stove. The mother goat takes her revenge when she invites her friend the fox together with the wolf for dinner. After dinner she invites her guests to amuse themselves by leaping over an opening in the floor; the goat leaps first, followed by the fox, but the wolf falls down into the fire and is burnt to death.

The other version of this same story has a happier ending, the goat challenging the wolf to leap a ditch where some workmen had cooked some gruel and left the fire still burning; the wolf tries and falls into the fire whereupon his belly splits open and the kids, still alive, skip out and run to their mother.

Another legend tells of a merchant who builds a house and sends his three daughters in turn to spend a night there in order to find out what they dreamed about. (Apparently there was a belief in parts of Europe that a man dreamed of by a maiden during the night of St John's Day, Christmas Day or the Epiphany was predestined to be her husband.)

The eldest daughter dreamed that she married a merchant's son, the second a nobleman and the third a goat, and in due course, despite precautions, she is carried off by a he-goat. Some geese bring news that her eldest sister is to be married and she asks the goat's permission to attend. He accedes to her request and also provides three black horses who transport her to the ceremony in three leaps. The goat, unknown to his wife, also goes to the wedding but in the form of a handsome young stranger transported on a flying carpet. The same thing happens at the second sister's wedding, but the third sister guesses the truth and leaves the wedding before the rest. On arriving home she finds the skin of the goat which she promptly burns, thus breaking the spell.

In Ireland the goat plays a prominent part in the 'Puck

Fair' of Killorglin, a ceremony in which fact and fiction are so interwoven that it is difficult to say where one ends and the other begins. It seems that up to the time of the Norman Conquest a notable feature of Irish life was the gatherings, resembling somewhat the clan gatherings in Scotland, at which every facet of Irish life, including industry, art, sport and literature, was displayed. When the English penetrated into Ireland efforts were made to suppress the gatherings, and the one which took place at the river-crossing where Killorglin now stands was prohibited. On the appointed day everyone observed the ban except a goat, which made its appearance on the usual site. The appearance of the goat was held to have kept the event alive and subsequently the custom developed of honouring a male goat in the now traditional manner of crowning and enthronement with the title of His Majesty King Puck. When

Goats on the Wicklow Mountains in 1955

exactly this custom was first adopted is not known, but there was certainly a fair associated with the ceremony as far back as 1613, when it was held at Lammastide, ie the first day of August. The fair now lasts for three days from 10 to 12 August and Margaret Murray describes the scene as follows:

> The Puck, after whom the Fair is called, is a male goat. This is a half-wild animal, living on the hills, and is caught for the sole purpose of presiding at the festival. Originally the privilege of providing the goat for the ceremony was vested in one family, though in recent years this has not always been the case.
>
> The first day of the Fair is called Gathering Day. Crowds surge to and fro through the streets of the town and the alleys of the Fair, drinking and making merry. The Market Square is the centre of attraction at all times. Later in the afternoon but before sunset (about 5.30 pm) the Procession of the Goat begins. This consists of a Pipe-band, followed by a lorry on which is the Puck-goat securely roped to a small platform. The Puck is decorated with wreaths round his neck, and is attended by four young boys dressed in green. After parading round the town for an hour the band and the lorry return to the Square, where a light scaffolding, 35 feet high, has been erected. A little girl, dressed as a queen with a crown on her head, crowns the goat with a tinsel crown and casts a wreath of flowers round its neck. Then the goat, still securely lashed to its platform, is hoisted with ropes and pulleys to the top of the scaffolding, where it remains till the close of the Fair. When the goat has reached its elevated position, a man proclaims through a megaphone 'The Puck King of Ireland'. At all other times the goat is referred to as 'The Puck King of the Fair'. Food of the kind beloved by goats is hoisted up at intervals, so that the animal is overfed during its captivity.
>
> The second day sees the festival at its height. The scenes, though now modified to drunkenness only, show that in early times this was one of those orgiastic festivals so common in primitive cults.

The third day is Scattering Day. The goat is lowered and set free, only to be caught up again, if possible, to be the Puck King in the following year. . . . The name Puck is a derivative from the Slavonic word Bog, which means God.

Scottish folklore includes a number of stories concerning goats and human fugitives, the best-known concerning Robert the Bruce who, fleeing from his enemies in 1306, hid in a cave at Inversnaid. Some wild goats lay down at the entrance and his pursuers, thinking that Bruce could not be inside, passed on. The king subsequently issued a decree that the goats should never be molested. A slightly different version of this story is that Bruce sheltered in a cave among the wild goats who provided him with shelter and sustenance in his hour of need. Whatever the truth may be it seems that a charter was granted by Bruce preserving the goats and delimiting an area of sanctuary called Pollochthraw (spelt phonetically) near Inversnaid wherein they were not to be harmed.

A similar tale is told of a gentleman who had taken part in the rebellion of 1715 and, after the battle of Preston, took refuge at the home of a relative in the West Highlands. It being judged unsafe for him to stay in the house, he went to a nearby cavern where an obstacle obstructed his entrance. He drew his dirk but was unwilling to strike lest he might take the life of a fellow in seclusion; and, stooping down, he at length discovered a goat and her kid lying on the ground. He soon saw that the animal was in great pain, one of her legs being broken. He bound it up with his garter and gave the goat water and bread, later venturing from the cave to gather grass and branches of trees, and the goat became much attached to him. It happened that the servant entrusted with the secret of the retreat fell sick and another was sent with provisions. The goat who happened to be

65

lying near the mouth of the cave opposed his entrance butting him furiously until the fugitive interposed and the faithful goat permitted the stranger to pass.

Early beliefs in the scape-goat possibly account for the superstitions about goats as removers of evil which remain in country places today. At one time it was the custom to bring a goat into the house where a sick person lay so that it might carry away the infection. Goats were also supposed to frighten away the baneful elf or goblin who, we are told by the Ettrick shepherd, frequented many Scottish hilltops and was believed by some to be the cause of the 'swelled head' from which their sheep suffered if not driven down to lower ground before nightfall. 'There's no luck where there's no goat' is another old saying, and the Tynesider would say:

'Good luck'll gan wi' ye, and yer kye'll de canny
As lang's yer kind te the aad nanny.'

This probably referred to the practice of keeping a nanny in the cow byre. On many farms even today farmers keep a goat with their other stock because they consider it healthy for the cattle; one of the maladies claimed to be averted by the presence of the goat is contagious abortion. It is said that abortion is often caused by the presence of ergot on the rye in the pastures, and that goats will consume this without ill effect and hence remove the danger. It is also claimed that goats, especially billies, eat the 'cleanings' and prevent them spreading abortion germs over the grazing. This superstition is not confined to Britain, for keeping a goat with cattle is frequently practised on the Continent also.

In Merionethshire it was thought that it was the smell of the goats that provided the barrier against sickness, and when the goat died the skin hung up in the cowhouse would still prove effective. Another theory of some Welsh farmers was that the goat's urine on the ground helped to prevent

abortion. The smell of the billy goat is also said to keep
rats away, whilst if run in the field it is claimed that a billy
will help keep cows together, and is also capable of putting
courage into a flock of sheep even to the extent of standing
up to a dog.

There is no doubt that goats like company whether it be
human or animal. A Scottish deerstalker once had a goat
which was very devoted to him and became a nuisance. It
had to be locked up before he went to the hill, but often
got out and joined the stalking party, where being light-
coloured and in the habit of bleating loudly, it was obviously
unwelcome. It also had a habit of standing under the pony
when a deer was being loaded on to the saddle and wiggling
its ears on the pony's belly. The pony would leap and kick
and on several occasions knocked the goat flying but
strangely enough never did it any serious harm.

Henry Tegner recalls an instance of a sheep farmer on
the Cheviots who reared a wild kid on the bottle. 'The goat
quickly learnt that the loose stone walls which surrounded
the field were easily climbed over. Soon after he became
friendly with the horse, he got out of his field and followed
the farmer on his pony out on to the hill. After that there
was no keeping the goat in. They were an unusual sight, this
trio of man, horse and goat, as they travelled over the moors
together.'

Sometimes goats are kept with horses who, it is said, love
the smell of them and help to keep them contented. It is
also believed that their odour prevents disease in horses and
there is an old saying:

> 'The hors'll always be healthy and able
> Where a goat is kept in the stable'

In the hunter stud farm at Birdstall the late Lord Middleton
always kept goats with the horses. His theory was that
horses used to goats never kicked hounds, and this proved

67

correct. A goat named Gin was the stable companion of Sandy Jane II, the Irish Grand National entry and another well-known Irish racehorse, Schmiss, also had a goat as a stable-mate.

A goat is supposed to be useful in the event of fire in the stable as he will walk out without panic while the horses follow, but such was not the experience of Mr S. Watney, of the brewery firm, who in 1938 obtained three goats for his company's stables in London. The goats did not take kindly to London life, continually escaped into Victoria Street and once caused chaos by eating most of the employees' time cards. When early in the war the stables were set alight by incendiary bombs the 85 horses were led away without trouble but the goats appeared panic-stricken and it was with difficulty that they were caught and carried out.

In the old 'Queen's Regulations' it was recommended that if a horse being embarked should refuse to pass along the gangway, a goat should be led before him; and the 'Judas Goat' is often employed at docks and abattoirs to lead sheep and oxen to the slaughter.

There are a number of beliefs among shepherds that goats are beneficial on the hill, some of these possibly arising because the goats tend to keep to the high ground away from the damp hollows where such things as fluke are more prevalent. 'He that harbours the goat will never have the poke' runs an old proverb—a swelling or poking of the skin under the neck of the sheep often being a sign of ill condition. A similar proverb runs:

'Rot nor poke nor loupin' ill,
I'll no come where there's a goat on the hill'

Many feral herds of goats owe their origin to the belief that on precipitous ground goats will keep the less nimble sheep out of trouble by feeding on the vegetation on rocky faces in preference to the more level ground which is left to the sheep. Commander J. R. C. Montgomery believed that the goats on Kinnabus Farm, Islay, had been introduced for the purpose of keeping the sheep off the cliff edge but they were a distinct failure since he spent a great deal of time rescuing sheep from the cliffs. Worse than merely inefficacious were the goats that a farmer, many years ago, put on to Pavey Ark, on Langdale Pikes, to keep the sheep off the rocks. They kept the sheep out all right but when one got on to the crags the goats promptly butted it off!

Beliefs connecting goats and snakes are not confined to Britain as in Kashmir the superstition that goats are deadly foes to snakes also exists and the name of the wild goat called the Markhor signifies 'snake-eater'. According to Bartholomew, writing in the sixteenth century, 'Serpents hate and flee the wild goat, and may not suffer the breath of him'. Buffon observes that goats are 'sucked by the *viper* and still more by a bird called the *goat-sucker* which fixes on their paps during the night'. Many of the old naturalists believed that the nightjar took milk from goats—hence its alternative name of goatsucker or goat-owl.

A belief that if goats 'lick serpents after these have cast their skin, they will not grow old, though they become white' is recorded in *Hortus Sanitatis* (c 1500), and Topsell reports that 'the wild goats of Egypt are said never to be hurt by scorpions'. Goats are said to kill adders by scooping them up with their horns and trampling them with their hooves, subsequently eating the body, leaving only the head. Hence the Gaelic proverb 'Cleas-na gooiths githeadh na nathrack', —literally, 'Like the goat eating the serpent'.

Other proverbs connect goats and adders: 'Goats kill edders and fatten the wedders' or:

'Goats kill ethers and help the yewes,
There's never na rot where the nannies browse'
are two sayings still extant in some country districts.

69

Opinions vary as to the effectiveness of goats in controlling adders and whilst from one locality in the Cheviots it is reported that adders are no less frequent in the area inhabited by goats than elsewhere, it is claimed that a certain grouse moor was so overrun by adders that it was too dangerous to shoot it with dogs until goats had been liberated on the ground. There is no doubt that adders, like most snakes, resent disturbance and the presence of numbers of large hoofed animals may well be inimical to them. Certainly some of the feral herds of goats existing today or formerly, owe their origin to the idea; the goats at Torrachilty for example are known to have been introduced in about 1880 by the local farmer to keep down adders.

Wild goats are said to be good weather prophets and to act as guides to the hill sheep during and before the advent of bad weather:

> 'Let him who would be weather wise,
> Mark well where the goat has lain:
> Before good weather he's up in the skies,
> But he comes down the hill for rain'

Another somewhat similar verse tells us that goats may be looked for:

> 'Up i' the hill for fine weather
> Doon i' the field for snaw;
> When it's dry they'll lie i' the heather
> On the rocks if it's gan te blaw'

Even the activities of the kids have something to tell the shepherd for 'if the kids are skipping, it is sure to be rain afore lang'.

Many of the early naturalists' observations on the goat have since proved to be 'unnatural history' but nevertheless make fascinating reading. Alcmaeon records that goats breathe through their ears and other writers have accused Aristotle and Topsell of perpetuating this story although Aristotle flatly denies it and Topsell leaves 'every man to his own liberty of believing or refusing'. A different method of inhalation is suggested by Oppian who believed that the goat breathed through his horns, but omits to tell us how the hornless breeds took breath!

Bartholomew considered that goats saw equally well by night or day but *Hortus Sanitatis* records that 'the goat does not see well in the daylight; but its sight is more accurate by night. The eyes of the goat shine by night, and they throw out light. Also he-goats have more teeth than she-goats'. The last belief is echoed much later by Buffon who was of the opinion that 'the number of teeth in the she-goat is not uniform: they are generally fewer than those of the male'.

Describing the horns of the wild goat, Turbervile, whilst fully aware that these were not cast each year like those of deer, did believe that the age wrinkles were shed annually. The same author gave goats credit for climbing 'marvelously for theyr feede', but 'sometimes they fal, then can they not hold with their feete, but thrust out their heads against the rockes and hang by their hornes untill they have recovered'. At other times, particularly during the rut when 'his throte and necke is marvelous great' even his horns could not save him from a fall but 'he had such a propertie that although he fal tenne poles length downe from an high, he will take no hurte thereby'.

Less fortunate was the kind of goat called by Turbervile the Ysarus or Saris which would sometimes 'skrat his thyghes with his foote, and thrusteth his hoofes in so farre, that he cannot draw them backe againe, but falleth and breaketh his necke, for his hoofes of his feete are crooked, and he thrusteth them farre into the skinne and then they will not come out agayne'.

The stink of the goat, especially the billy, is proverbial and the Reverend Bingley relates the following story as to the origin of the smell. 'Goats are exceedingly numerous in South Guinea; and some of the negroes there have a singular notion that their strong and offensive smell was given to them, as a punishment, for having requested of a certain female deity, that they might be allowed to anoint themselves with a kind of aromatic ointment which she used herself. Offended at the request, they say, she took a box containing a most nauseous compound, and rubbed their bodies with it; and that this had so powerful an effect as to cause the unpleasant smell thence produced to continue ever afterwards'.

Also proverbial is the lust of the goat and many early writers remark on the short life of the domestic billy due to his ardour for female company. Topsell writes that: 'There is no creature that smelleth so strongly as doth a male Goat, by reason of his immoderate lust, an in imitation of them the *Latins* call men which have strong breaths (*Hircosi*) Goatish . . . An therefore *Tiberius Cesar* who was such a filthy and greasie-smelling old man, was called (*Hircus vetulus*) an old Goat.'

Bewick remarks on the excessive appetites of the male goat and remarks that one buck is sufficient for one hundred and fifty females, while Mascal suggests that after five years a billy could no longer serve a nanny whilst the latter, at the age of eight would become barren. 'There is no beast that is more prone and given to lust than is a Goat, for he

Goats in flight on Bray Head, Co Wicklow in 1955

joyneth in copulation before all other beasts. Seven days after
it is yeaned and kidded, it beginneth and yeeldeth seed . . .
although without proof. At seven months old it engendereth
to procreation, and for this cause that it beginneth so soon,
it endeth at five years, and after that time is reckoned un-
able to accomplish that work of nature. That which is most
strange and horrible among other beasts is ordinary and
common among these for . . . the young ones being males,
cover their Mother, even while they suck their milk.'

5

The Uses of the Goat

The goat was probably the first ruminant to be domesticated by man and plays an important role in many early cultures. In the course of time it tended to be ousted by the sheep, especially in temperate zones, doubtless because the sheep has better flesh, more fat and produces wool; while with the coming of domestic cattle the importance of the goat as a milch animal declined. However, the goat is able to live in places where sheep cannot thrive, especially in arid mountainous areas where grass is scarce and thorny scrub predominates and today it is more important than the sheep in many countries such as Africa, where it flourishes in the most inhospitable places.

The parable of the prodigal son is one of many biblical references to the goat which shows that the kid was the ordinary dish at an entertainment, the fatted calf being reserved for some special occasion. Goats formed an important item in the wealth of the patriarch being used for sacrifices as well as food: among the presents sent by Jacob to propitiate his brother Esau were two hundred she-goats and twenty he-goats; Nabal possessed one thousand goats and goats formed a considerable portion of the possessions of Laban and of the wages of Jacob.

The skin of the goat has been used from time immemorial for the conveyance of water, wine, milk and oil, and skin

Goats on Bray Head, Co Wicklow in 1955

bottles are repeatedly alluded to in the Scriptures. Tristram gives the following account of their manufacture:

The animal is skinned from the neck by simply cutting off the head and legs, and then drawing the skin back, without making any slit in the belly. The apertures for the legs and tail are at once sewn and tied very tightly up; and the skin, in this state, with the hair on, is steeped in tannin and filled with a decoction of bark for a few weeks. There are large tanneries in different towns, where the process is carried out on an extensive scale; especially at Hebron, where bottle-making, both of glass and leather, is the staple of the place. The skins are there partially tanned, then sewn up at the neck, and filled with water, the surfaces being carefully pitched. They are then exposed to the sun on the ground for several days, covered with a strong decoction of tannin and water pumped into them from time to time to keep them on the stretch till sufficiently saturated.

The hair was left on the bottle, helping to preserve the skins from damage although in time it wore off and old ones could be recognised by their baldness. Pegler mentions that inflated skins were used as swimming bladders, in particular by Assyrian fishermen around 884 BC.

The goat was an important animal in the economy of ancient Egypt from the predynastic period onwards, the skin being used for wrapping the dead, and there is a reference in Exodus demonstrating the use of the hair— 'And thou shalt make curtains of goats' *hair* to be a covering upon the tabernacle'.

The skins of goats, especially kids, were formerly important for the making of parchment, this use being reflected today in the name of the strong all-rag paper known as 'Goatskin Parchment'. According to Rutty the skin of the very young kid serves for 'making Fans, Screens, Kites, and etc. for which purposes the external lamina of the Hair side is sometimes stripped off, being very fine, and not apt to crack as Paper.' The skin is also, observes Pennant, 'well adapted for the glove manufactory, especially that of the kid; abroad it is dressed and made into stockings, bed-ticks, bolsters, bed-hangings, sheets and even shirts. In the army it covers the horseman's arms and carries the foot-soldier's provisions. As it takes a dye better than any other skin, it was formerly much used for hangings in the houses of people of fortune, being susceptible of the richest colours; and when flowered and ornamented with gold and silver, became an elegant and superb furniture'.

The Rev W. Bingley described the method of tanning goats' skins as follows:

76

There is a way of preparing them by maceration, so as to separate the fine upper pellicle, or epidermis, from the coarse under parts; after which they are dyed of various colours for different uses. From the skins of Goats is manufactured what is generally called *morocco leather*. The countries most celebrated for this are Turkey and the Crimea. In the latter, only the red and yellow morocco are prepared; but these, in point of quality, are fully equal to those of Turkey. The best morocco is made of he-goat's skins.

The same author, quoting from Anderson's *Recreations in Agriculture,* drew attention to the underwool present at the roots of the long hair of some breeds of goat, which, while neglected in England, was extensively used in Russia for manufacturing gloves, stockings, etc on which a high value was placed.

Angora and Cashmere goats are the best-known breeds which produce wool in economic quantities. Cashmere wool is the fluffy undercoat of a breed living at high altitudes in the Himalayas, the wool being combed out directly from the goats once a year. The breed will flourish in Britain where, however, the climate is too mild for the production of a good fleece.

The fine, silky hair of the Angora goat is known as mohair and first became an article of commerce in 1749 when a few Dutch and English merchants, who had settled in or near the town of Angora, commenced buying mohair and spinning yarn for export. It was not, however, until about 1836 that the spinning of mohair became an industry in England. Pegler comments:

The chief use to which mohair is applied consists in the manufacture of Utrecht velvets, generally known as furniture plush, and largely used in France, Germany and the United States for the linings of railway carriages, also for sofas, chairs and table-covers. A great proportion of this Utrecht velvet is made in Amiens in France, the mohair being previously spun at Bradford, the centre of the mohair trade in England, which indeed supplies the mohair yarn for the whole of Europe.

Angora goats will also live in Britain but the rainfall is too high for the production of good hair, although David Mackenzie mentions that the late Duke of Wellington kept a flock in Hampshire which 'produced mohair of a quality comparable to the second and third grades of imported mohair'. The breed is extensively farmed today in the USA, South Africa and, to a lesser extent, in countries of the eastern Mediterranean.

In the eighteenth century the long hair of goats, especially the males, was in demand by peruke makers for their best and whitest wigs. The best hair for this purpose came from the haunches and it was baked and bleached before use: according to Pennant, 'a good skin well haired, is sold for a guinea, though a skin of bad hue, and so yellow as to baffle the barbers' skill to bleach, will not fetch above eighteen-pence or two shillings.'

Goat hair was also fabricated into ropes, such as those used by the hardy natives of St Kilda to swing themselves over the steep cliffs of their island in search of sea-bird eggs. In water it is claimed that a rope of goat hair will last longer than one of hemp. It has been suggested that the goat's beard was the original of the white sporran worn by some of the highland regiments and long goat hair is still used for sporran-making. Another use for a goat's beard was as a cider-strainer.

The flesh and fat of the goat have been esteemed from the earliest times—in Deuteronomy both goat and wild goat were included among the ten 'beasts which ye shall eat'. Buffon, however, was of the opinion that goat was never as good as mutton regardless of how it was raised, except 'in very warm climates, where mutton is always ill tasted'.

Pennant records that goat afforded a cheap and plentiful meat for the inhabitants of the more mountainous regions of Wales. 'The haunches of the goat are frequently salted and dried, and supply all the uses of bacon: this by the natives is called *Coch y wden,* or hung venison'. He goes on to say that 'the meat of a splayed goat of six or seven years old (which is called Hyfr) is reckoned the best; being generally very sweet and fat. This makes an excellent pasty; goes under the name of rock venison, and is little inferior to that of the deer'.

In Ireland, Rutty states:

The flesh of the male goat, castrated and fed, makes a good venison and lately kids are reared about the mountainous part of the south of Dublin for the delicacy of the flesh preferable to that of the lamb: for this purpose they are taken into the house presently after they are dropt, viz: before they have tasted their mother's milk, and suckled by ewes and fed with cow's milk spouted into their mouths.

Goat horns have been put to many uses, as drinking vessels, horns, knife-handles and hat racks, and even fashioned into bows by having the tendon of an ox stretched across them.

The tallow of the goat was generally agreed to be good for making candles; indeed both Topsell and Pennant thought that candles made from the fat of goats better than those made from sheep or ox suet, and the latter author mentions that wild goats were hunted in the autumn in Caernarvonshire to provide fat for this purpose.

There has, of course, always been a use for goats' milk which, provided it is produced under proper conditions, is indistinguishable by most people from cows' milk and is in fact more easily digested—or as the older authors put it, 'not so apt to curdle upon the stomach'. Unfortunately goats' milk was accused of causing cases of brucellosis in Britain early in the present century, and although the charge was completely unfounded—in fact it was cows' milk that was at fault—the good reputation of goats' milk was never properly restored. David Mackenzie, who presents a well-argued case for the production and use of goats' milk, laments the fact that today it is associated with stone-ground flour, compost-grown carrots, anti-fluoride hysteria, and acupuncture. 'However sound the arguments in favour of back-to-nature enthusiasms, they are expressed too often and too vehemently by Colonels' daughters nostalgic for a lost prestige.'

Goats' milk also makes excellent cheeses and the Cheddar district of Somerset produced its famous cheese originally from the milk of sheep and goats and not from that of cows. A cure for some respiratory ailments was to sit inhaling a rotting cheese made from goats' milk, and goats' cheese is also said to appease all wounds and pains if laid upon them. Pennant extolled the milk of goats as sweet, nourishing and medicinal; if mixed with a teaspoonful of hartshorn and drunk warm in bed in the morning and again at four o'clock in the afternoon, and repeated for some days, it was a good cure for 'pthisical people, before they were gone too far'.

At one time the crofters living on some of the west-coast islands of Scotland had to subsist, when bread was short, on a dish called *oon* which, according to Martin (1703), was made up as follows:

A quantity of Milk, or Whey is boyl'd in a Pot, and then it is wrought up to the mouth of the Pot with a long Stick of Wood, having a Cross at the lower-end; it is turn'd about like the Stick for making Chocolat, and being thus made it is supp'd with Spoons; it is made up five or six times, in the same manner, and the last is always reckon'd best, and the first two or three froathings the worst; the Milk or Whey that is in the bottom of the Pot is reckon'd much better in all respects than simple Milk. It may be thought that such as feed after this rate, are not fit for action of any kind, but I have seen several that liv'd upon this sort of Food, made of Whey only, for some Months together, and yet they were able to undergo the ordinary Fatigue of their Imployments, whether by Sea or Land, and I have seen them travel to the tops of high Mountains, as briskly as any I ever saw.

Apparently it is immaterial whether *oon* is made of cow or goats milk. If the latter is used it 'is thickened and taste much better after so much working; some add a little Butter and Nutmeg to it'. Comparing *oon* with chocolate, Martin considered that 'if we judged by the Effects, this Dish was preferable to Chocolat, for such as drink often of the former, enjoy a better state of Health, than those who use the latter'.

In the eighteenth and early nineteenth centuries the drinking of whey made from goats' milk was, as mentioned earlier, a minor health fad and many feral herds trace their origins from goats kept to supply this demand.

Rutty had quite a bit to say about whey drinking—the proper season for which, he said, was March to May, by reason of the fact that the nanny kids in March. Whey drinking was, however, 'continued by many in June and July, but even in June the Milk thickens, for which reason they then mix four ounces of water with a quart of Milk before they turn it, and more water in July, the Milk growing thick as the season advances, so that in August it is not to be drunk; but in September is a second Spring and the Milk becomes thinner again and may be used medicinally, tho' not with equal advantage as the former season'.

As a foster-mother the goat is invaluable, and many a young life, from lion cub to foal, has been saved by being reared on goats' milk. In Western Australia a goat once acted as foster-parent to a young camel, the huge youngster having to kneel down to take milk from the nanny goat.

As a source of fresh milk during journeys, before the advent of the refrigerator, goats have been unrivalled from ancient until quite recent times; as Low puts it, 'goats bear well the motion and confinement of shipboard, and are better fitted for supplying milk on sea voyages than any other animal.'

Various parts of the goat were held by the ancients in very high esteem for their medicinal virtues, and even up to the sixteenth and seventeenth centuries such organs as the liver, gall, spleen and horns, not to mention the blood, milk and excrement, were all being used as remedial agents. During an epidemic or pestilence a goat was sometimes sacrificed to the rising sun; Pegler recalls an incident when 'the people of Cleonae, an ancient town of Argolis, did this, and when they were freed from the plague, they sent a brazen figure of a goat to Apollo'.

Goats' blood is recommended as a sudorific, aperient and resolvent, useful in treating pleurisy, retention of urine, and stone in the bladder; the ancient Greeks believed that boiled goats' blood was an effective antidote for poison. Most medieval writers attributed remarkable powers and solvent properties to goats' blood, which included dissolving diamonds, destroying venom and curing baldness. Topsell claims that 'the bloud of a Goat hath an unspeakable property, for it scoureth rusty iron better than a file, it also sofeneth an Adamant stone', and also records the effectiveness of the blood of the wild goat against dizziness: 'there are Hunters which drink the bloud . . . coming hot out of his body . . . against that sickness'. Van Helmont affirmed that

> If you hang a goat by the horns, bending the hind feet to the sides of his head, and in this posture cause his testicles to be cut out, and then dry the blood that runs from the wound, it will become as hards as glass and difficult to

beat into a powder. One drachma of this will infallibly ease and cure the pleurisy without blood-letting.

Other ancient uses for goats' blood included restoring the magnetic property of the loadstone, shattering false emeralds, engraving on glass and crystal and smearing the palms of the hands of sleeping men in order to extract their secrets. It was also useful for calling up witches, who, however, were quite capable of revenging themselves upon having their rest disturbed, for on their Sabbaths they were in the habit of murdering goats, black ones—and eating them raw. Indian magicians who drank the blood of a goat sacrificed to the goddess Kali were believed to be temporarily possessed by her spirit and thus enabled to prophesy. Pliny tells us that magicians used the flesh of a goat roasted at the fire wherein a man's body is burned, against the falling sickness (a malady which Topsell believed could be contracted by eating the liver of a goat which had not been properly prepared for the table) while goats' fat was said to have astringent qualities and was applied with salt to whitlows or felons, and dropped into the ear for deafness.

The supposed antipathy between goats and snakes crop up again in the belief of ancient writers in the effectiveness of the hair, burnt wool, and ashes of the horns in driving away serpents; the hooves, burnt and pounded with liquid pitch, cured baldness. Insomnia could be infallibly cured by placing the horn of a goat under the head of the patient and vineyards rendered fruitful by burying horns points downward in the soil at the rootstock of the vine to conduct rainwater to the plant.

The belief that goats could see as well, or better, by night than by day doubtless gave rise to the use of the liver and gall in helping night vision and dimness of sight. The liver was also considered a good remedy for hydrophobia

83

and the gall mixed with honey for dropsy or with bran for dysentery. Pliny also mentions that the dung of goats, as well as the ashes of the hair, 'expelleth the stone'.

Goats' milk has long been used medicinally and Topsell mentions its use in preserving wine against sourness: like asses' milk it has long been thought to have cosmetic properties and there is a Gaelic saying which translated runs:

With violets and the milk of goats anount thy face freely,
And every king's son in the world will be after thee (my dearie).

In recent times goats' milk has been, and is, used widely in the treatment of infantile eczema, duodenal ulcers, asthma and other complaints.

Another potentially highly useful attribute of a goat product, one which appears to have fallen into disuse, is related by Topsell who tells that the urine of the goat mixed with spikenhard may be drunk as a cure for love.

6

The Chase of the Wild Goat

George Turbervile gives a vivid description of the chase of the wild goat prior to the sixteenth century. After remarking that he had not seen or heard of any wild goats in England, he goes on to say that there may well be some in Wales and in other mountainous districts, and since it appears from the holy scriptures that the flesh is venison it would be as well to give an account of the way that goats should be hunted:

> The best time to hunt the wilde Goate, is at Alhallowtide, and the huntsman must lie by night in the high moun-taynes in some shepeheardes cabane, or such cottage: and it were good that he lay so seven or eight dayes before he meane to hunte, to see the advantages of the coastes, the Rockes, and places where the goates do lie, and all other such circumstances: and let him set nettes and toyles, or forestallings, towards the rivers and bottomes, even as he would do for an Harte . . . if he have not hewers nor Huntesnen ynow to set rounde aboute: then let him place his companions on the toppes of the Rockes, that they may throw downe stones, and shoote with Crossebowes at the Goates . . . This chase requireth no great Arte nor following, neyther can a man follow on foote nor on horsebacke.

Both Cox and Pye quote freely from Turbervile but neither of them say whether the method of hunting the goat that he describes was actually practised in Wales, or anywhere else, during the eighteenth and nineteenth centuries.

Bartholomew mentions how the goats of his day evaded capture, however, by throwing themselves off the high crags and saving themselves by landing on their horns. Also current at this time was the belief that goats and deer when wounded went in search of the herb dictamnus which, when eaten by the stricken animal, ejected the arrow from the body and healed the wound. This herb was so called because it grew plentifully on Mount Dicte in Crete—the English name being dittany.

But for the fine horns carried by some of the older billies few sportsmen would bother to kill a feral goat, for the meat of the adult animal is not really fit for human consumption. Their horns will, however, always attract the trophy-hunter and even the most experienced stalkers can have an attack of goat fever, when for the first time they see these splendid horns outlined, perhaps, against some rocky skyline. That much-travelled sportsman, the late Frank Wallace, was no exception:

> He was standing with his head up, every inch of his splendid horns thrown into strong relief against the pale lilac of an evening sky, and as I watched him that attack of goat fever began, which in the end nearly made me lose him, and for the best part of three days rendered me the most unhappy man in Sutherlandshire.

Unhappy he must have been, for on the first day the sight of those horns was too much for him and he missed twice; two days later he again found himself within a hundred yards of the big billy, broadside on and in a fair light, but missed again, and it was not until his next attempt that he finally succeeded.

It was John Colquhoun who, over a century ago, first suggested the feral goat as a suitable quarry for the rifle. ' I have often thought ' he wrote, ' that for those who have

Head from Burren Hill, Co Clare: length 39¾in, span 36in

a taste for deerstalking, without the opportunity, it might be no bad substitute to have a flock or two of goats upon a remote range of hills.' One of the reasons that prompted this suggestion was his knowledge of a fine old billy which broke away from a tame herd in Glen Douglas, remaining at large for almost a year and becoming so completely wild that it took half a dozen shepherds, ranging the mountains with guns, some days to shoot him.

A few owners of forests did introduce goats for stalking, one of these being a Mr Bainbridge who, in 1900, introduced about twenty-five white goats to Achnashellach. After a few years on the hills they became exceedingly wild and afforded excellent sport. On some forests, which already had a stock of feral goats, the owners introduced ibex to improve the heads, one being the late Sir Keith Fraser, who about 1903 introduced a pair of ibex to Inverinate with, it is said, very good results. Early this century the eleventh

87

Duke of Bedford, who rented Cairnsmore in Kirkcudbright-shire, turned out two male hybrids between the domestic goat and the Grecian ibex, but generally when an owner considered new blood necessary it was the male goat that was introduced.

That great sportsman, author and artist J. G. Millais had a very high opinion of the sport offered by the Scottish wild goat:

> The lover of the rifle who has a month to spend in June or July, when other sport with the rifle, except roe stalk-ing, is scarce, might do worse than pass his time in obtaining a couple of good goat heads. Permission to shoot them is often easily obtained, and the sport is excellent. When they have been pursued for several years goats soon learn to take care of themselves, and even if their chase is not quite so difficult as that of other hill game, the hunter can at least pursue them by himself and they will teach him several wrinkles in the art of stalking.

From a sporting point of view it is not difficult to go out and shoot a wild goat, but to kill any particular head is usually a far different proposition: the same is true in deer stalking. Millais gives an admirable description of the best type of sport that could be expected in his account of his pursuit of a wild billy that roamed the mountains near Garve in Ross-shire. It was an immense white beast and for a whole week he pursued the herd without once getting within shot of it, but on the last day, which dawned fine after a week of rain, Millais was out early and by 5 am found the herd grazing on a steep shale slope:

> Two old females kept guard, as usual, whilst the males fed and the kids played about. The position the goats were in was in the hollow of a corrie and one that would make a successful stalk nearly impossible, so I lay for three hours on the hill above waiting for them to move. At nine they galloped out of the corrie and went nearly

two miles to the east, and then, after feeding for a while, all lay down except one female, who stayed on watch. The position of the goats was now scarcely more favourable than before, so I took up a position about 500 yards above them and waited nearly the whole day for them to move. At four they left their position and began to ascend the hill towards me, and then turned west again, as if making for the ground where I had found them in the morning. The two old males, however, lagged far behind and as I kept parallel with the moving herd the females gave me an opportunity to ' cut in ' down hill. This I succeeded in doing by taking advantage of a dry watercourse. As soon as the last of the herd of females had passed below, I made all haste downhill and got into a position which I hoped would be well within shot of the lagging old billies as they came into view. After waiting for five minutes the horns of the black male showed up below and to the right just at the spot where I had expected, and presently the owner came walking quickly along the narrow path. He looked a fine beast and had I not seen the big white one I dare say I should have been glad to secure his horns. But I was looking for something better, so he was allowed to pass on. I had rested another five minutes, there was no sign of the white one, and I feared he had passed me using another road below. Accordingly I descended, and taking the cover of a large rock looked in the direction in which I expected to see the object of my desire. He was, however, nowhere to be seen. He had, in fact, gone by and then ascended abruptly and joined the herd under the hill beyond my vision.

As I knew he would not willingly leave the herd I again ascended the mountain, and after walking half a mile to the west, circled so as to cut the line of the herd of females. I observed them again at 5-30 feeding slowly along below me, and accompanied by the black male. Concluding that the white one had either lain down or was still concealed in one of the numerous depressions above which I had passed, I then retreated on the line the herd had come, spying carefully every hollow. As

I crossed one open space something moved on the sky-line, almost on the level with myself, and looking intently I discovered it to be the horns of a goat. The next moment the patriarch stepped into view and stood on the ridge. The evening sun shone on his snow white coat and glistening horns and a gentle breeze waved his mighty beard. He certainly looked finer than many a truly wild beast, and I longed to possess those spreading horns. Now he looked straight at me, and being detected fairly in the open one could do nothing but keep quite still and hope for the best. I hid my face in the heather and did not look up until a tinkle of stones warned me that game was on the run. As good luck would have it, however, the goat did not retreat beyond the ridge, in which case he would have been out of sight at once, but ran along it and then down into the dip, attempting to pass below. The shot was a long one, about 150 yards, but I sat up at once and had ample time to achieve a good shooting position. At the first shot from the Mannlicher the goat stumbled and turned straight down-hill, the bullet having hit him too low, but at the second he collapsed and was quite dead when I got up to him. This was a very fine specimen. The horns were of the divergent type, 34 inches long and very massive.

If left undisturbed, goats are fairly regular both in habit and habitat and this greatly facilitates the search. Goats, however, are cliff-loving animals, and whilst approach may be comparatively easy, as often as not the chance of a shot may be nullified by the fact that the animal is standing in such a position as to rule out any reasonable chance of re-covering the trophy. To shoot the head which, perhaps, for several seasons the hunter had viewed with longing eyes yet spared for improvement, and then see it go crashing several hundred feet down some scree or cliff face to finish up, perhaps, in the sea, would indeed be a mortifying experience. For first-class heads are always few and far between, and deserve all the care and patience that would be devoted to a ' royal ' stag.

I well remember my first introduction to feral goats. I was photographing birds on the island of Ailsa Craig and had just reached the highest point of some rocks which stood back from the cliffs proper, when a most repugnant goaty smell intruded upon the sea-fowl odours around us. Sure enough a quarter of a mile further on we saw a small herd of nine goats slowly walking away from us. In a matter of seconds they had disappeared from view, but not before the impression of an immense spread of horn, sweeping back and outwards, carried by the leader kindled the desire to acquire such a trophy for my collection. All thought of bird photography was abandoned as we endeavoured to stalk the big billy and get his portrait; eventually I succeeded by concealing myself on a likely pass whilst my companion slowly moved the herd toward me, and the resulting picture only confirmed my first impression of a great head.

It was not until some ten years later that I was again to find myself in goat country, not with a camera this time, but with a rifle and the kind invitation of the late Lady Bullough to shoot a good head on the island of Rhum.

It was late one January afternoon when I reached the island to be greeted by the head keeper, Duncan Mc-Naughton, who assured me that there were perhaps a hundred goats on the island and felt sure that I would go back to England satisfied. He had never before been out after a goat and remarked that he thought I would find the goats easy stalking after the stags. How far from the truth did that remark turn out to be !

Our plan was to spend two or three days viewing the beasts and selecting our quarry, leaving the remaining days to be spent trying to bring him to account. The first day was bitterly cold, with driving rain to dampen our ardour and fog up our glasses; nevertheless we saw between forty and fifty beasts, and in particular one shaggy black billy with outward-curving horns which seemed very nearly to confirm my memory of that first billy seen on Ailsa Craig. He was peacefully sitting on a narrow ledge some 300 feet above the Atlantic, and although we could approach unseen to within about 150 yards, to shoot him where he lay would have been tempting providence. Next day, although we saw a further thirty or so goats, none was the equal of this billy, so plans were set to spend the remainder of my' time trying to bag him.

The third day we soon spotted him feeding with two other billies amongst the seaweed at the foot of the cliff; here was the ideal spot to get him for there would be no danger of him falling over the rocks. But before we had completed half of the perilous descent he and his companions had forsaken the shore and were on the cliff face, at about the same height as ourselves, slowly walking away. For the rest of that day we followed them, scrambling over loose boulders and rocks, crawling along ledges and climbing up and down the cliffs until we were utterly exhausted. In the end I did get to within about fifty yards of him, surprising him as I slid around one of the abutting prominences. For fully two minutes he stood on his narrow ledge, glaring back at me with his yellow eyes. A shot would have been easy, but I should only have seen his shaggy form bounce from ledge to ledge to the rocks below. Then with a shake of his head and a sharp hiss, he moved off in that tantalising walk of theirs which appears so slow, yet rapidly increases the distance from you. And that was the last I saw of him that day.

The following day was passed in similarly futile chase but we did spot a billy with a slightly smaller head, and as he was in a favourable position Duncan suggested that I should take him in case we failed to come to terms with the big one. He was standing with three others on the cliff top, about ten yards from the edge and there was ample cover for an approach to within about eighty yards. When I reached the selected point, however, all four goats were slowly walking along the cliff top, but as he was broadside and gave a reasonable target, I fired. At the shot he gave a convulsive leap forward and then disappeared over the cliff to drop some three hundred feet to a rocky beach. When, by a devious route, I reached the carcase, it was obvious that the head, smashed to pieces by the fall, would be useless as a trophy. It was a timely lesson that should I have a chance at the big one on the morrow, no shot should be taken unless the beast was well clear of the cliff top.

On my final day the stalker, despite a twisted ankle, valiantly agreed to take to the hill again, so with a full flask of pre-war Scotch, we set forth just before daylight to the spot where we had left 'William' the previous evening. Mist hung heavily about the cliff tops and at times completely blanketed our vision. After about an hour, through one of the wreaths of mist swirling along the cliff face, Duncan spotted our goat. Experience of the past few days had taught me that it was practically hopeless to try to stalk him from the rear, so I selected a spot, well in advance of the direction in which he was feeding where there would be a reasonable chance of recovering him, and awaited his pleasure to pass my way.

The only place which appeared suitable was a narrow gully which ran steeply down the face of the cliff and appeared to finish up on a broad slab of rock just above the water. Extreme caution was necessary to reach the desired spot without dislodging any of the many boulders which strewed the descent, or clicking my metal studs against the rocks, for either would have given warning of my approach and possibly alarm some nannies which were feeding half a mile further on.

I had barely reached the selected point before a bearded face with yellow eyes and wide-spreading horns appeared on the edge of the gully; after a short look round he hastened to cross its stony bed, for it was no use lingering there in search of food. When he was directly below me, drawing a bead on the side of his broad, shaggy shoulders I fired. He dropped to the shot, at first sliding and then rolling, he bounced with alarming momentum down the gully and disappeared from view.

On reaching the body we found to our delight that he had landed on a narrow ledge a few feet above the water, his head quite unscathed. The gully, which from the top appeared to run out on to a slab of rock, in fact terminated with a 20 ft vertical drop to the ledge where we now found

Billy killed on the Isle of Rhum in 1944

him—the slab of rock was separated from the mainland by a strip of surging water. The fates had indeed been generous.

That afternoon 'William's' head, destined for the taxidermist and wrapped in canvas, accompanied me on my train journey south. For the greater part of the distance I had the compartment to myself, but at one small station a gentleman carrying a number of parcels joined me. By now I was practically immune to goaty smells and I wondered how the stranger would react. My surprise may be imagined when, after opening the window and sniffing the air, he anxiously enquired, ' I hope you don't object to the smell of herring ? I have just bought some for the wife and I'm afraid they're smelling the carriage out." It takes a good smell to keep another down !

7

Enclosed Goats of England

In addition to the various herds of feral goats inhabiting the British countryside, goats have also been included in collections of animals kept in some English parks, though few remain today. Three hundred years ago goats were kept in Tredegar Park, Monmouthshire; in *The Life of Marmaduke Rawdon of York* (Camden Society, 1663) we read:

> Tredegar, a seate belonging to Squier Morgan . . . He hath there a stately parke through which runs the river Ebwith, soe that in his parke he hath salmon, trouts, and what fish that river doth afford . . . He hath a warren near his parke, and in his parke a thousand head of deere, besides wild goats and other cattle about his grounds, soe I think he is pretty well provided towards house keepinge.

At one time there were goats in Calke Abbey Park, near Ticknall but a billy buffetted a child which subsequently died and the goats were killed, the offender's head being displayed in the keeper's house.

Goats obtained from Cairnsmore Hill, Kirkcudbrightshire, were formerly kept at Woburn Abbey the last one being disposed of about 1953, and other parks, which at the end of the last century harboured goats, wild or otherwise, included Easton Park, Woodbridge in Suffolk (50 goats); Arundel in Sussex (herd of Cashmere goats); Caley, Otley, Chevin in Yorkshire; Lowther near Penrith, Westmorland

(21 goats) and Ingestre near Stafford (26 goats). In the British Goat Society Year Book of 1920 a letter describes a herd of wild white goats, apparently of ancient origin living in an unspecified park in Herefordshire.

Perhaps the most famous of the park herds was the herd of Cashmere goats which, for over a century, roamed the Great Park of Windsor Castle before being disposed of in 1936 when the majority of the animals were deposited in the Regent's Park Zoo where their descendants still remain. This herd, no longer pure, forms the source from which many of the Welch Regiment mascots are obtained.

J. Wentworth Day gives an interesting account of the origin of the Windsor herd:

> It was in the autumn of 1823 that Christopher Tower, the squire of Weald Hall near Brentwood, in Essex, was staying in Paris when a large herd of Cashmere goats arrived there from that fairytale principality in the valleys of the Himalayas because the French had an idea that they could be acclimatised in France, and a profitable industry established in their wool. Before that time, as many as 16,000 looms had been constantly at work in the State of Cashmere turning out many thousands of shawls a year.

> Mr. Tower, like all his family, was a keen naturalist, a born traveller and a man of vision. If the French could do it, he could do it. So after some difficulty, he bought two pairs of goats from this imported herd and shipped them over to his park in Essex. They stood the change of climate well and soon there were lots of Cashmere kids gambolling among the sunny glades of Weald Park, greatly to the dignified consternation of the enormous red stags who, for centuries, had been the monarchs of these oaken glades.

> Since Cashmere goats produce only about 4 oz. of their very fine wool in a year and the nanny only half that weight, it obviously took Squire Tower some little time to collect enough fine wool, sort it out from the coarse long

hair and have it spun and woven into a single shawl. However, in 1828 a shawl was produced, Mr. Tower won the Gold Medal of the Society of Arts, and King George IV was so impressed by his achievement that, in June of that year he gladly accepted a pair of goats from him.

That was the beginning of the Windsor herd. The pair soon multiplied and when Queen Victoria came to the throne she, and later the Prince Consort, took a close personal interest not only in the goats but in the manufacture of Cashmere shawls. Shawls were very much in the fashion in those days and it is said that over a thousand people were engaged in producing them from the Windsor herd. Many were ladies who volunteered their services for the slow and difficult process of separating the fine down from the coarse hair.

In 1874 a pair of the Windsor goats were exchanged with the Duke of Buckingham for a pair of his Cashmere goats from Stowe Park. The Duke, incidentally, had received eleven goats in 1863 from his uncle, the Marquess of Breadalbane, of Taymouth Castle in Perthshire. Despite this introduction of new blood, however, the royal herd was suffering badly from inbreeding, so in 1889 a fresh herd was sent from India as a present to the Queen, to introduce new blood and fresh vigour into the Windsor stock. Alas, they were nearly murdered with mistaken kindness on the voyage. The goats had brought with them not only a strong and vigorous strain of new blood but an equally strong and vigorous strain of Himalayan lice and fleas. Some well-meaning ship's officer soaked them liberally with an anti-bug mixture which consisted mainly of paraffin oil. This, coupled with the hot sun and sea air on the voyage, killed every hair on their body. The goats arrived at Windsor stark naked.

The transfer to London Zoo was not quite the end of the Windsor goats for one old nanny got left behind and for many years was to be seen in close company with the red deer. Still alive in 1951 when the herd of deer was disbanded, she must have been very lonely during the last years of her life.

As for the original Weald Park goats, a small herd still existed there up to at least 1923 but on the outbreak of war in 1939 the park became a military camp, the fine herd of deer was either butchered or escaped and the house itself, partly gutted by fire, was pulled down shortly after the war.

Less famous but no less interesting is the herd of some twenty goats kept in an enclosure near Blithfield Hall, Staffordshire—the ancestral home of the Bagot family. These goats are the remnants of an ancient herd that roamed Bagot's Park and wood, its history going back as far as the fourteenth century. The goats were formerly kept within the 800 acres of the park but during the period 1937-57 had access also to Bagot's wood and the surrounding countryside, the park fences having fallen into disrepair. When at large they seldom wandered much further afield

Goat Lodge near Bagot's Park. Note the stone frieze of goat heads

than the wood, which, together with the park, gave them a range of about 2,000 acres, much of it woodland where a few fallow deer also lead a precarious existence. In 1957 the Forestry Commission took over the woods and it became desirable to restrict the movements of the goats lest they damaged the young trees. Accordingly some 110 acres of park and woodland were fenced off for the goats and there they remained until the death of Lord Bagot when all but a dozen or so were sold to Mr R. Bagot of Levens Park near Kendal. A few evaded capture and found refuge in the wood adjacent to Goat Lodge, later purchased by Mr Phil Drabble, the naturalist. These animals were subsequently added to the small herd remaining at Blithfield which is now transferred during the winter months to the wood at Goat Lodge, returning to Blithfield in April each year to be on show to the public who visit the Hall in the summer.

Compared to the other herds of wild goats in Britain the colouring of the Bagot goats is unique in that the head, neck and shoulders is jet black, whilst the rest of the body is pure white, and to keep the breed as pure as possible all

Schwarzhals goats in Bagot's Wood in 1950

badly marked goats are killed. It is the colouring of this herd that makes these goats so interesting for it is none other than that which typifies the Schwarzhals breed of the Rhone Valley.

This goat, known in its native land as the Glacier goat or Saddle goat, comes from the Canton Valais and is especially numerous in the Rhone Valley. It is long-haired and always has horns, those of the male being long and often widespread, whilst those of the nannies rise close together and are shorter. The height at the shoulder of a typical billy is about 30 in.

In common with other feral herds the billies at Blithfield generally go to rut in the autumn and the majority of kids are born from late January to the end of March, although an occasional kid may be born during any month of the year.

The history of the goats is tied up with the history of the Bagot family, one of the oldest in Britain and one of the few to own land prior to the Norman Conquest. Major-General the Hon George Wrottesley suggests that the Bagots were a Breton family, as the name in its original form of Bagod and Bagoth has a very celtic appearance and the field of the early Bagot shields was always ermine—the well-known coat of Brittany. The name has also been variously spelt Bigod and Bigot.

About 1380 a goat's head appeared for the first time on the crest of Sir John Bagot. Why he selected a goat is not recorded, but from then onwards the goat has figured prominently on the Bagot coat of arms; in the family church at Blithfield the crest is carved on the tombstones of Lewis Bagot (1461-1534) and Thomas Bagot, his son (1504-41), while over the tomb of Richard Bagot (1552-96) and his wife hangs a helmet which has, mounted on the top, the head of a goat.

There are various theories regarding the origin of the

Richard Bagot's helmet hanging above his tomb in Blithfield Church

Bagot goats: John Rodgers suggests they were brought over from Normandy at the Conquest, while a former Lord Bagot believed they were presented to a predecessor, one John Bagot, by Edward II for services rendered, possibly on one of the crusades. The late Lord Bagot discounted the latter suggestion and believed the goats did not reach Bagot's Park until the end of the fourteenth century.

In 1385 the Duke of Lancaster (John of Gaunt) formed an alliance with the King of Portugal with a view to obtaining possession of the crown of Castille, which he claimed in right of his wife the daughter of Pedro the Cruel. The following year, accompanied by an armed force and some of his supporters, he set sail for the Con-

102

tinent, including in his party one Sir John Bagot. It is possible that when the Duke's crusaders returned to this country in about 1387 they came overland up the Rhone Valley, home of the Schwarzhals goat, possibly bringing with them some of the goats as souvenirs or, more likely, as a source of milk. Moreover, Sir John Bagot, who had probably received his knighthood shortly after the coronation of Richard II in 1377 was the first Bagot to use the goat's head on his crest and it may be that when he saw the Schwarzhals goat in the Rhone Valley he was so impressed with their appearance that he decided to bring some home to his estate, if only because their presence there would fit in appropriately enough with the family crest.

Another popular theory is that the goats were presented to Sir John Bagot by Richard II during the early part of his reign, in acknowledgement of a good day's hunting in Bagot's Park, and if the presentation had taken place before John Bagot received his knighthood it provides a plausible explanation for his selection of a goat for his crest. That still leaves unexplained the problem of how King Richard obtained the goats in the first place, but it is possible that they were brought back by Crusaders returning home on foot through Europe. If either of the last two theories is correct then the goats first came to Bagot's Park between 1377 and 1388 and have been there for 600 years. Most of the old English families have some legend or other, and it is not surprising to find that one of those of the Bagot family predicts that when the herd dies out, so will the house of Bagot. Obviously, therefore, succeeding Lords of Bagot have gone to considerable trouble to ensure the preservation of their herd!

Over 250 years ago a Sir Edward Bagot was compelled, in the interest of agriculture, to reduce the number of goats, but the reduction was only temporary and at a Court

A well marked Schwarzhals billy in Bagot's Park in 1954

at Stafford in 1710 Sir Edward received permission to increase the herd ' to an unstinted number.' At the outbreak of World War II, however, the fate of the goats, which then numbered about eighty, hung in the balance, for the War Agricultural Executive Committee, with the same ruthlessness with which they tackled the deer problem in many other parks, decreed that the entire herd should be destroyed on the grounds that the goats were damaging crops. The Lord

Bagot of the day successfully pleaded their case and permission was granted to keep the herd at not more than sixty, which number was maintained until about 1947 after which the herd increased gradually to a hundred. In 1954 the herd was again reduced to sixty by the elimination of badly marked animals, most of which were killed although a few were sent to Wales, subsequently to be released on the Rhinogs in Merionethshire, while some of

During 1938 the late Lord Egerton of Tatton Park, Cheshire, received eight of the Bagot goats which were reduced to a pair in the severe winter of 1940. Some years later, solely by inbreeding, their numbers had increased to eight again and in 1952 there were eleven, by which time most of the goats were more brown and white than black and white. There are, however, no longer any goats remaining at Tatton Park.

The herd purchased by Mr R. Bagot in 1962 has survived and numbered twenty in 1969, although about half of these had very little black marking on their bodies.

Gazetteer of Wild Goats

*Maps showing localities where
feral goats exist or formerly
existed appear on pages 147-157*

ENGLAND AND ADJACENT ISLANDS
(*maps on pages* 151, 152 *and* 154)

Note: The number in brackets against each locality refers to its location on the appropriate map.

Devon

Lundy (178)

Goats abundant certainly from 18th century but by 1914 the last survivors of a large herd of white goats were killed. Goats re-introduced in 1926 to supply milk for lighthouse keepers soon went feral; increased to about 200 whereupon some were killed and numbers decreased to about 90 in 1952.

Lynton Rocks (176)

Goats on Lynton Rocks descended from those placed in the Valley of Rocks many years ago by the owner.

Rough Tor (177)

A few on Rough Tor near Okehampton earlier this century are now extinct.

Gloucestershire

Kingswood (166)

Goats numerous in Old Forest of Kingswood in 17th century, but have long been extinct.

Isle of Man

Calf of Man (149)

Goats on Calf of Man believed descended from those kept by lighthouse keepers 1818-75.

Isle of Man (148)

A few in the Sulby Glen-Snaefell and Laxey Glen area where goats have existed for many years; head of a billy goat shot in 1905 on Cronk ny Earey Lhaa (north of Port Erin) in Manx Museum, Douglas.

Lancashire

Coniston (144)

Several herds on the fells between Coniston and Tilberthwaite; herds of up to 40 near Coniston village survived into early part of this century. Last survivors seen on Yewdale Crag and Holme Fell about 1915.

Northumberland

Callerhues (141)

Just north of the river North Tyne at Bellingham there was for many years a herd on the Callerhues on Blakelaw.

Cheviots

Goats split up into small herds mostly fairly local; individuals or small parties may travel long distances and have been reported from Hareshaw, on the Brigg Fell to Nunwick Moor; Catcleugh and Cottonshope in Redesdale and Harthope Linn on the southern face of Cheviot.

Cheviot and *College Valley Herds* (139)

Herd of mostly blue-grey goats north of the Cheviot mountain and east of the College burn; brought to valley about 1860 apparently to replace an older herd existing at Southernknowe which was removed elsewhere. Numbers reduced from 28 to 14 in the 1946-7 winter, increased to 32 by 1950 when all but 9 were shot for damaging crops but now (1971) recovered to about 30.

Christianbury Crag (138)
Christianbury Crag herd on Bewcastle Fells near the
Cumberland border, between 30 and 40 in 1950s; goats
seen before World War II on South Tarset moors near
Ealstone possibly strays from this herd.

Kielder (135); *Plashetts* (136)
Until recently two herds in the Kielder district centered on
Plashetts and Christianbury Crag. Former frequented
the Earl's Seat, Wainhope and Monkside region of
Plashetts, north of the River North Tyne. Goats which
may belong to this herd are sometimes seen in upper
Redesdale, near Catcleugh reservoir, ranging south to
Emblehope Moor or north to Hungry Law. Plashetts
herd 60 in 1955 but most were killed for damaging
Forestry Commission plantations at Kielder; the few
survivors may now be extinct.

Thrunton Crag (140)
East of the Cheviots the remnants of a small herd exist at
Thrunton Crag, south of Whittingham near the Bridge
of Aln; 3 or 4 in 1962.

Whickhope and *Simonburn* (137)
Simonburn goats in two herds, one in Townshiels farm
district (perished in winter of 1947) the other at Goat-
stones about 4 miles west of Simonburn. Latter herd also
suffered in storms of 1947 and reduced to a few nannies;
a billy from Plashetts was introduced but by 1959 only
4 or 5 animals remained. Whickhope herd dates certainly
from 17th century, about 50 at end of last century but
many killed when tree-planting commenced in 1940s and
the few survivors perished during 1946-7 winter.

109

Somerset

Athelney (175)

A reference from 10th or 11th century mentions goats near Athelney (Somerset), but possibly roe deer may be meant (Greswell, 1905). Early this century a herd of Welsh goats said to be running practically wild on the cliffs of North Somerset. No goats in county today.

Westmorland

Martindale (143)

Goats mentioned by Wordsworth had disappeared by 1805; never numerous in Lake District and apparently none there today.

Yorkshire

Kilnsey Crag (146)

Small herd of mostly brown and white goats on Kilnsey Crag near Arncliffe in Upper Wharfedale; about 21 in early years of this century; now extinct.

WALES
(map on page 154)

The majority of wild goats in Wales today is found in the north, particularly in Caernarvonshire and Merionethshire where their numbers have been recently estimated at 300. Afforestation has caused the extermination of some herds and the outbreak of foot and mouth disease in 1957-8 resulted in the thinning of the goat population to reduce the risk of spreading the infection.

Anglesey

Amlwch (151)
A few goats reported (1961) from cliffs on north coast west of Amlwch.

Caernarvonshire

Glyders (152)
Most goats in Snowdonia (153) are in the rugged country between Llyn Ogwen valley and Pass of Llanberis, and range between Glyder Fawr, Glyder Fach and Tryfan: wanderers on Y Garn, Elidyr Fawr, Y Foel Goch, Moel Perfedd and Snowdon. Total population (1964 estimate), 59 nannies; 19 billies. Goats 70 years ago in Nant Gwynant and Mynydd Mawr near Beddgelert were shot for damaging walls. North of the Glyders, goats seldom cross Llyn Ogwen valley and are rarely seen on Carnedd Dafydd, Carnedd Llewelyn or Foel Fras. Herd of about 60 white goats near Bettws-y-Coed exterminated early this century.

Great Orme's Head (150)
Herd, not truly wild or Welsh, derives from Cashmere stock introduced from Windsor Great Park about 80 years ago and kept mainly to supply mascots for Royal Welch Fusiliers; joined by local strays the herd numbered about 25 in 1953.

Moel Siabod (155)
Herd of 30-40 early this century; probably now extinct as much of the area has been afforested. In 1951 a goat ('one of a herd of 27') was caught on Moel Siabod but these probably wandered in from Merioneth.

Mynydd Mawr (154)

About 8 goats during the inter-war years; by 1952 only 3 remained of which 2 were known to be billies.

Tremadoc Rocks (157)

Herd which existed during last century extinct by 1900.

The Rivals (156)

Herd (about 30 in 1961) of grey-and-white or roan goats has frequented The Rivals (Yr Ei fl) and Nant Llithfaen district since last century; their future problematical as area is being afforested.

Cardiganshire and Carmarthenshire

Goats formerly existed on Plynlimon Fawr range (separating Montgomeryshire from Cardiganshire) and on the Black Mountains which lie between Carmarthenshire and Brecknock.

Merionethshire

Cader Idris (161)

Goats formerly found on Gesail (near Machynlleth) and the Cader Idris ranges on north side of which lies Llyn-y-Gafr (Goat Lake).

Craig-y-Benglog (159)

Goats along north side of road between Dolgelly and Llanuwchllyn; introduced about 1870 and semi-domestic, ownership claimed by farmers; range from eastern slopes of Rhobell Fawr and Dduallt in north to rocks of Wenallt just above main road. Nannies (about 20) mostly on Craig-y-Benglog; billies prefer Craig-y-Ronwydd but visit nannies in summer and stay for Autumn rut.

Moelwyn (158)

Herd (which existed over a century ago) now numbers 15-20 mostly black-and-white goats which in hard weather may come down to foothills in Maentwrog valley and Tan-y-Grisiau rocks just above Blaenau, and occasionally wander on to Cvnicht.

Rhinogs (160)

About 50 goats in mountains near Llanbedr range extensively over Craig-y-Saeth above Cwm-Bychan Lake, sometimes on Rhinog Fawr, Rhinog Fach, Foel Ddu and lesser peaks. When disturbed may wander to Yllethr and Diphwys; occasionally souih to Llawr Llech near Barmouth and north to the Y Graigrddrwg range. In 1957 5 goats (derived from the Bagot herd) released on the Rhinogs by Major F. Bennett. The story that a herd of Afghan goats was imported for the film *The Drum* (shot at Cwm-Bychan just before World War II) and liberated on the hills is apparently groundless.

Pembrokeshire

Fishguard district (162) and *Prescelly* (163)

Goats that frequented cliffs near Fishguard and, inland, the Prescelly hills now apparently extinct; the white goats on Dinas Head destroyed by Pest Officers in 1942.

Radnor

Craig Pwll Du (164)

Early last century a herd existed in this wild ravine near confluence of the Bach Howey and Wye; it is not known how long it persisted.

113

SCOTLAND: MAINLAND
(*map on page* 148)

Although plentiful on some of the islands off the west coast wild goats are not numerous on the mainland of Scotland except in the south west. Small herds are, however, present in most of the northern and central counties.

Aberdeenshire

Bennachie (34)

Small herd (6 in 1932) on Bennachie, about 8 miles west of Inverurie; now extinct. Goats formerly seen in Glenbuchat, near Alford, probably wanderers from Bennachie.

Glencallater (35)

Herd originated about 1850 by John Gordon of Auchallatar; about 80 in 1939, only 7 by 1949 and probably extinct by 1959.

Angus

Glenisla (113)

Small herd on cliffs of Creagan Chaise originated about 25 years ago with stock from Auchavan Farmhouse. Joined in 1946 by a billy (probably from Glencallater); by 1952 only the billy, a nanny and kid remained. By 1959 only 1 nanny left—perhaps the kid referred to above.

Argyllshire

Ardmaddy (77)

On the west coast opposite the Island of Seil there were goats in the 18th century near Ardmaddy but they have been long extinct. According to Mr H. Mortimer Batten there used to be a herd on Goat Island, Loch Craignish which may still exist.

Ardgour (72)

Herd (about 50 in 1937) of dark brown or black goats at Ardgour (north east of Morven), by 1968 reduced to about half; Gaelic name is Aird-Ghobhar (the Goat Rock). At the lower end of Glen Gour is Lochan Gabhar (Goat Loch) but there are no goats there now.

Benderloch (75)

East of Loch Linnhe there were (1937) indefinite reports of goats on Benderloch and (about 1914) in the hills around Loch Creran at about which time a billy was shot at Barrs on the west shore of Loch Etive—possibly having wandered in from Glen Coe. In Loch Linnhe goats formerly present on the Isle of Lismore and on Shuna Island reduced by 1964 to 2 nannies.

Conaglen (71)

North of Ardgour goats on the Conaglen estate less plentiful than formerly, have their headquarters on the Guesachan ground at head of Loch Shiel; occasionally wander on to Ardgour. About 20 seen (1958) at head of Loch Eil.

Glencrippesdale (69)

In Morven 2 small herds on the Glencrippesdale estate have their headquarters on the hill above Camas Saluch; joined about 1940 by liberated domestic stock. Goats formerly on island of Carna (entrance to Loch Sunart) now extinct. Early this century 50 goats on Aonach Mor or Innean Mor or Morven but they have been long extinct.

Glendaruel (101)

Herd in Glendaruel usually found in the wild country just north of Glen Striven, apparently exterminated by Forestry Commission about 1940.

Glen Hurich (70)

Goats formerly on Forestry Commission ground at Glen Hurich the progeny of domestic stock left by a shepherd in early part of this century; their numbers at first increased and then declined until introduction by Lord Morton of 2 billies, when they began to increase again. During World War II a number shot and by 1951 only about 2 dozen remained; herd is now extinct; last one shot about 1956.

Kilmalieu (73)

30-40 goats on Kilmalieu estate east of Glencrippesdale live chiefly on the ground between Meall Na Each and Druim Na Maodalaich facing the sea (Loch Linnhe); sometimes seen on adjacent Kingairloch. Mostly dark grey with a few black and white but in 1952 one with typically Schwarzhals markings.

Mull of Kintyre, Carradale (98)

Goats on Mull of Kintyre for many years. Pennant mentions them in 1772 and Dugald Macintyre (1952) states that the caves of the Mull of Kintyre, of Jura and Islay and of Mull and elsewhere in Argyll have been the winter shelters of goats for ages past. Apparently there was a herd on Island Davaar and some may still survive there. Goat population of the Mull still considerable especially in the south-west around Rudha Duin Bhain, Uamh Ropa and Earadale (about 50). Goats sometimes seen near the lighthouse and probably roam most of coast between Carskey Bay and Machrihanish. On east side of the Mull (some 14 miles north of Campbeltown) about 40 white goats on Carradale Point for many years; locally believed to be descended from goats put off a ship about 150 years ago, not very wild and probably descended from a domestic Saanen stock. Early this century there were some exchanges with other herds of similar type; about 1949 2 kids were sent to Flichity, Inverness-shire.

Tighnabruaich Forest (100)
Forestry Commission reported (1964) about 6 outside of the forest which lies some 8 miles south of Glendaruel, facing island of Bute.

Ayrshire

Glen App (118)
Herd probably originated from imported Irish goats which strayed or were abandoned during last century; died out about 1920.

Kirrieroch Hill (117)
Goats still seen around Kirrieroch Hill and Mullwharchar, just north of the Merrick on Kirkcudbright border.

Dunmfries-shire

Craigieburn (129)
Just north of Beattock and Moffat a herd (about 20) formerly frequented the Craigieburn; killed when Forestry Commission took over the area, except for a young pair which moved on to Roundstonefoot Farm where they bred and were joined by strays from Blackshope. By 1952 there were 4 billies, 4 nannies and 2 kids. Further up the glen there are goats at Saddleyoke, Birkhill and around the Grey Mare's Tail which sometimes join the Roundstonefoot herd; total number in the area about 50. Grey Mare's Tail herd frequently found on Polmoody Farm, sometimes wander up the Moffat Water into

Selkirk but seldom cross into Peebleshire by way of the Carrifran and Blackshope glens; sometimes seen on lower slopes of Bodesbeck Law on south side of Moffat Water. Herd around Birkhill, probably never more than 30, reduced to 3 after winter of 1947 but increased again to about 12 after a stray billy (probably from Langholm) wandered in.

Langholm (133)

Near Langholm (east Dumfries) a herd (about 20) for at least 40 years—their headquarters around Tarraswater, especially a ravine near Arkleton Hill; wander west in Eweswater and seen near Mosspeeble and as far as Megget Water, Eskdale.

Queensbeery Hill (128)

Small herd early this century on Queensbeery Hill (6 miles south-west of Moffat) became extinct probably during the 1920s. In 18th century goats numerous near Raehills. A few goats seen recently between Thornhill and Water of Ac.

Dunbartonshire

Inch Lonaig (102)

Goats once inhabited the yew-tree island Inch Lonaig on Loch Lomond, causing considerable damage to trees. John Colquhoun (1841) writes that they were ' very large and the oldest inhabitant does not recollect when they were first introduced '.

Inverness-shire

Achdalieu (54)

According to McConnochie (1923) goats on Achdalieu

forest were almost exterminated by 1835 but in more recent years as many as ' a dozen bucks have been seen together.' East of Lochiel lies the Great Glen where, in the 18th century, Dr Johnson encountered great herds of goats. South of Loch Eil, goats on Conaglen (Argyll) can sometimes be seen from the road at the head of Loch Eil but are apparently prevented by road and railway from straying north on to the Locheil ground.

Affric (48)

Glen Affric lies to the north of Ceannacroc and the Forestry Commission reported (1959) a herd of some 40 goats, mostly outside Commission land.

Brin, Flichity (42)

Old-established herd on the Flichity estate in Strath Nairn formerly numbered about 50 but by 1944 decreased by half; by 1952 only a billy, a nanny and their 4-year-old son remained. These goats were pure white and had their headquarters around Brin Rock.

Ceannacroc (49)

Ceannacroc forest lies at the head of Glenmoriston and harboured many black and grey goats in the early years of this century; by 1952 they had all been killed or moved away.

Corrour (55)

Goats on Corrour forest, between Glen Spean and the Blackwater reservoir, killed off between 1891 and 1897.

Dalmigavie (44)

East of Glenkyllachy the herd on Dalmigavie numbers about 20.

119

Dochfour (41)

Herd locally believed to be descended from animals belonging to men who built the Caledonian Canal. The goats frequent the moors above Dochfour House especially the highest point (Red Rock) and also go south to the Abriachan moor: herd before War World II about 100 strong but almost wiped out by Canadian troops during the war; now about 20. A white billy introduced about 1925 from Brin and a billy from Mull about 1937. West of Dochfour the herd in Strath Glass near Struy, apparently extinct.

Eilean nan Gobhar (58)

' Isle of Goats ' in the Sound of Arisaig locally believed to have supported a herd of goats during the time of Prince Charlie (1745-6). This herd has long been extinct but during World War II a goat from south of Inverailort was deposited on the island by commandos.

Farr (40)

North and east of Flichity no goats remain on Farr or Glenkyllachy; herd on former died out early last century.

Glenmazeran (43); *Coignafearn* (45)

In the central Monadhliath mountains a herd on Coignafearn deer forest which marches with Dalmigavie and Glenmazeran in the upper Findhorn. During the 1920s there were 60-80, reduced by 1944 to about 40. Estimate for 1968 for the area suggested 150 goats.

Glenmoriston (47)

Herd on Glenmoriston hill in 1956 numbered 16; exterminated by Forestry Commission by 1958.

Inverailort (57)

No goats now resident; wanderers occasionally seen.

Invergarry (50)

In 1885 a Mr Macdonald of Invergarry bought 12 nannies and a billy at a sale of wild goats at Carrbridge; some went to Morar when Mr Macdonald moved but what happened to the rest is not known; no goats at Invergarry today.

Kinveachy (37)

East of Dalmigavie the herd (26 in 1951) on the Kinveachy forest has existed for over a century.

Loch Hourn (51)

A few goats in the mountains bordering the loch and formerly on Kinlochourn; wanderers also reported from adjacent forests of Glenquoich and Arnisdale. About 50 years ago a herd of dark-grey goats existed above Loch Hourn on Eilanreach; now apparently extinct, as are the goats which once frequented Rattachan. Goats also reported (1938) from a small island in Loch Hourn near the entrance to Glen Barrisdale. South of Loch Hourn wanderers are seen on Barrisdale and formerly on Knoydart and Glenkingie.

Loch Shiel (56)

A herd in the hills around Loch Shiel, now apparently extinct, the last animal shot about 1954.

Meoble (53)

Probably resident herds existed on the forest of Meoble during the last century but now only wanderers occur, some as far west as Arisaig. Millais (1906) commented on the fine domestic goats of this area: 'magnificent animals with horns measuring up to forty-six inches.'

North Morar (52)

South of Loch Nevis an old-established herd of about 20 (mostly grey) goats on North Morar deer forest; usually found on the south side of Sgor na h-Aide and wander on to Carn Mor. In 1900 a Mr Macdonald brought goats (probably from the Monadhliath Mountains) to Morar, some of which joined the local herds.

Port Clair (46)

South of Glenmoriston herd of about 20 goats on Port Clair deer forest exterminated about 1930.

Across Loch Ness from Port Clair a herd on Beinn a Bhacaidh died out before World War II.

Rothiemurchus (36)

Goats formerly in Rothiemurchus, more precisely the Queen's forest of Glen More, since the early 19th century; originally apparently two herds, one around Craiggowrie north of Loch Morlich was exterminated shortly after the Forestry Commission acquired Glen More, by which time (1923) the other herd near Allt Creag an Leth-choin was probably already extinct.

Slochd and Corrybrough (38)

North of Kinveachy herd of about 100 formerly on the North side of Slochd, often seen from the main road; reduced in the winter of 1947 to about a dozen—most of these subsequently poached. A few at Corrybrough wander about between the Findhorn river about Tomatin, south through Balnespick towards the river Dulnain near Carrbridge.

Upper Findhorn (39)

Where the river Findhorn crosses the boundary between Nairn and Inverness a herd (29 in 1949) often seen about Cairn Kincraig and Creag a Chrocain, some 3½ miles east of Loch Moy. Possibly originated from stock from the farm of Balchrochan, vacated in the 1920s.

Kirkcudbrightshire

Most of the wild goats of southern Scotland are in Kirkcudbrightshire, about 750 animals range over almost the whole mountain chain from Merrick and Rhinns Kells in the north to Cairnsmore of Fleet in the south. Wild goats occur as individuals, small parties or herds throughout the Galloway hills, but culling by the Forestry Commission has reduced numbers in recent years.

Craignelder (124)

Herd (about 40) on Craignelder Hill exterminated by Forestry Commission in 1952. In winter goats often seen around New Galloway Reservoir and Murray's Monument where they shelter in abandoned shepherds' houses, some of which have a bed of dung 5ft deep.

Cairn Edward Forest (125)

Herd (about 25) at Cairnsmore Black Craig of Dee exterminated by Forestry Commission but an occasional stray seen on Allendoch. In 1962-63 the Commission rounded up 122 billies from the Round Fell area which were subsequently killed.

Cairnharrow (126)

Herd on Cairnharrow (east of Creetown) probably died out about 1930.

123

Cairnsmore of Fleet (123)

Craignelder goats often join up with the herd (about 100) on Cairnsmore of Fleet. Early this century the 11th Duke of Bedford liberated 2 male hybrids between domestic goat and Grecian ibex but they have left no obvious mark on the present stock. Prior to 1914 some Himalayan Tahr were liberated on the hill but finally died out.

Dalbeattie (127)

Goats near Dalbeattie exterminated when Forestry Commission took over the area.

Glen Trool Forest (121)

This forest (and the 2 following) now owned by Forestry Commission. Moderate-sized herd often seen at Caldons, south of Loch Trool, has its headquarters south of Loch Dee near the White and Black Laggan burns. Also a small herd on the Buchan and Glenhead Hills north of Loch Trool. About 36 goats said to be on the Commission ground in 1959.

Kirroughtree Forest (122)

6 goats resident (1959) on Kirroughtree Forest, adjoining southern section of the Newton Stewart to New Galloway road. When forest was first opened up there were hundreds of goats on the neighbouring estate of Bargaly and large numbers were killed.

Merrick (120)

Before World War II about 200 on the Merrick and adjacent Benyellary Hills; only a few remain.

Lanarkshire

Elvanfoot (130)

Until recently a few in hills around Elvanfoot, Crawford; by 1958 only one billy remained.

Morayshire

Lochindorb (33)

The herd (about 30 in 1952) on Lochindorb Moor occupies about 2 square miles around the Glentarroch Rocks and derives from 7 survivors of a larger herd which left Dunearn Moor during World War II.

Nairnshire

Findhorn (32)

Goats have lived in the Findhorn Valley for at least a century. Today there are two main herds (one north and one south of the river) ranging over Dunearn and Drynachan (Nairnshire), Lochindorb (Morayshire) and Dalrachne (Inverness-shire). Goats north of the Findhorn (about 40 in 1968) are usually on the Cawdor estate. In bad weather they may come down to the woods above Drynachan Lodge and in the early months of the year frequent the Carnach Burn; in Spring they move west to high ground between the Cawdor and Moy Hall estates and are often seen on a hill called Marchlet and also on Cairn Kincraig and Creag a Chrocain near the Inverness-shire border.

Peebleshire

Peel (132)

Goats at Peel near Walker Burn, about 8 miles east of Peebles, destroyed when Forestry Commission acquired

Elibank. Only area where goats may still occur is extreme south west where they may wander in from Dumfries-shire near Hartfell Rig and near Stobo.

Perthshire

Ardnandave (107)

Formerly a herd of about 12 on Ardnandave Hill near Strathyre and a single black goat reported from Tulloch on the shores of Loch Voil. By 1959 both localities were reported free of goats by the Forestry Commission. Goats also reported in past from around Loch Lairig Eala and Glen Ogle.

Atholl Forest (110)

Goats listed among game driven to Queen Mary during her visit to Blair Atholl in 1564 and still present early in 19th century; now apparently extinct.

Ben Venue (105)

South of Lochs Katrine and Achray old-established herd on Ben Venue formerly numbered over 100; reduced to 30 by end of World War II and 8 by 1951.

Braes of Balquhidder (108)

From time to time, certainly since late 19th century, goats seen on mountains between Glen Falloch and the Braes of Balquhidder, including Beinn Chabbair, but seldom seem to stay long.

Glenartney (106)

East of Strathyre and the Braes of Balquhidder lies Glenartney deer forest. A petition (1763) by the tenants of Dalclathick claims reduction of rents because they had been forbidden to keep their goats on the forest as they had done formerly; however, goats appear to have remained until the end of the 19th century when the last was found dead on Stuic-a-chroin. In 1957 a billy (probably from Ben Lomond) appeared and remained until his death in 1964. In 1965 a pair was introduced from an island near Jura, followed by a nanny in kid from Ben Lomond. By 1968 there were 3 billies, 2 nannies and a kid.

Loch Ard (104)

Much of this area now forms the State Forest of Loch Ard and the herd (formerly about 30) on the slopes of Ben Venue has been culled by the Forestry Commission. By 1959 only a nanny and kid remained. To the west is Ben Lomond on which there is a herd and doubtless there has been some interchange between the Ben Venue and Ben Lomond stock as small groups have been reported at Corriegrennan about halfway between the localities.

Meggernie (109)

Herd (about 40) of mostly black goats on Meggernie, in Glen Lyon, disappeared during World War I. The highest mountain on Lochs which overlooks Glen Lyon, is Stuchd an Lochain where, in 1590, Mad Colin Campbell and his followers encountered a flock of goats, though there are none there today. In 1950 a single goat appeared on the forest of Dall (north of Glen Lyon).

Rohallion (112)

During last century Millais (1914) records a herd (about 15) that roamed the hills between Rohallion and Kinnaird; none in the area today.

Schiehallion (111)

Large herd on Kynachan hill, between river Tummel and Schiehallion killed during World War II to provide meat for Indian troops stationed in Inverness-shire.

Ross & Cromarty (mainland)

An Teallach (12)

Small herd (about 12) for many years on An Teallach Mountain (Dundonnell Forest). These long-haired, pure white goats still present in 1952 and occasionally strayed on to the adjacent Inverbroom Forest. Two dark billies seen (1958) on Creag Rainich on north side of Loch a Braoin. Recent estimate about 40 in area.

Achnashellach (22)

The 25 white goats introduced to Achnashellach forest (1900) apparently died out between the two World Wars; until 1952, however, a few wandering goats generally a bluish-grey and white in colour, were reported, but none in recent years.

Badrallach (8)

H. Boyd Watt (1937) mentions herd of about 40, owned by a shepherd but indistinguishable from feral animals, at Badrallach on the north shore of Little Loch Broom.

Beinn Eighe (21)

In 1910 a Mr Macdonald came to Taagan (a farm at the head of Loch Maree) bringing a dozen silver-grey goats from Morar. These soon went feral on Meall a Ghuibhais and were joined by 2 tame grey kids introduced in 1925. By 1952 the herd numbered about 50 mostly darker than those liberated. According to local opinion goats already existed on Beinn Eighe before these introductions.

Black Isle (31)

Before World War II there were 2 herds on the Black
Isle, one in the Culbo area (now Forestry Commission
property), the other (about 20) on coast between Rose-
markie and Cromarty. Both now extinct, the last-
mentioned shot by Norwegian troops stationed in the
district. Charles St John who lived in the Highlands
from 1837-53 reported a herd on cliffs of Black Isle,
possibly the ancestors of the herds mentioned above—he
also reported an old goat and 2 kids near Cromarty. ar
North west of Black Isle goats occasionally seen in the
Freevater Forest locality and in 1953 a billy killed on
Glencalvie.
Glencalvie.

Coigach (5)

Before World War II a large herd on the coast of Coigach
Rock facing Enard Bay; another herd (about 40) on
Ardmair hill just north of Ullapool exterminated during
or before World War II. About 12-15 seen on Inverpolly
Forest.

Dundonnell Hill (11)

On the north side of Strath Beg local goats introduced
for sport in 1911; by 1928 about 30 goats on Dundonnell
Hill. A few still left by 1952 often seen near road from
Dundonnell House to Fair. ı.

Fisherfield Forest (14)

Before World War II a resident herd; now only strays
from Gruinard and Letterewe are seen.

Flowerdale and *Shieldaig* (17)

A herd (about 20) on forests of Flowerdale and Shieldaig;
further south Torridon has no resident herd but strays

occasionally seen. Goats were formerly present on Liathach, the highest mountain on Torridon.

Glenshiel and *Clunie* (26)

Goats often seen on Glenshiel and Clunie; resident herd (about 10) on Glenshiel certainly until 1952. Goats also wander in from Kintail and 2 billies (one white and the other with typical Schwarzhals markings) have been reported from Clunie; a similar Schwarzhals-type billy appeared on Kilmalieu (Argyllshire) in 1952.

Glomach (28)

East of Inverinate and Kintail are the forests of Glomach and Affric (Inverness-shire); goats sometimes seen on former but none resident on Affric. About 1950 a herd (about 30) reported above the Glomach Falls, but the more usual haunts of goats in this area is A' Ghlas-bheinn on Dorusduain.

Gruinard Forest (13)

Adjacent to Dundonnell, Gruinard Forest has a long-established herd numbering about 50.

Inverinate and *Dorusduain* (24)

North of Loch Alsh and Loch Duich about 50 on Inverinate and Dorusduain deer forest have their head-quarters on each side of and between A' Ghlas-bheinn and Beinn Fhada or Ben Attow, the latter mountain forming part of Kintail forest. Before 1918 200-300 goats at Inverinate but were considerably culled during the war. About 1903 the late Sir Keith Fraser introduced some ibex (either *Capra ibex nubiana* or, more likely, the wild goat, *Capra hircus aegagrus*) which, it is claimed, inter-bred with the existing feral goats. In addition to the ibex some Irish goats were introduced about 1912.

Kinlochewe (20)

Originally Kinlochewe deer forest extended to both sides of Loch Maree, but the Beinn Eighe section is now Nature Conservancy property. North of loch there are goats on the Letterewe side of Slioch (page 132); herd (about 50) often seen around Loch Garbhaig and Lochan Fada on north side of Slioch possibly a collection of the herds from Letterewe. There have been goats on Slioch for many years, reinforced by a few of those brought by Mr Macdonald from Morar—though most of these went elsewhere (page 128).

Kintail (27)

Goats on Ben Attow and neighbouring A'Ghlas-bheinn are mentioned (page 130). On the Five Sisters range another herd (about 60) is based on Sgurr na' Carnach and Sgurr na Ciste Duibhe facing Glenshiel, sometimes seen by the roadside at Shiel. Last time goats were liberated here was 1880 when animals belonging to a Miss Peggie Mary Macrae were turned loose; ruins of Miss Macrae's cottage near Clachan Duich Churchyard still visible.

Lael (10)

Part of Inverlael forest (Lael forest) now owned by Forestry Commission but goats in the area apparently increasing in numbers; about 40 in 1959.

Leckmelm (9)

South of Ullapool there were goats on Leckmelm before World War II but they are now scarce; some of those seen are wanderers from a small herd on the Ardcharnich hill-ground which often range as far as Mheall Dubh on the march between Leckmelm and Inverlael.

Letterewe Forest (19)

Three herds on Letterewe Forest; one (about 15) on south side of Slioch, near the Kinlochewe Forest march; another (about 20) on north and south slopes of Beinn Airidh Charr and the third (about 30) on south side of Mhaighdean and Sgurr ne Laocainn. These goats all probably originated from Gruinard and their colouring is similar. Opposite Letterewe a few on Eilean Suthainn (largest island on Loch Maree) introduced by a Mr J. MacRae have spread on to mainland around Talladale.

Rattachan (25)

West of Glenshiel goats once plentiful on the (now) Forestry Commission property of Rattachan are now extinct.

Red Point (16)

West of Shieldaig goats frequent coast along north shore of Loch Torridon from Diabaig (herd of about 16) to Red Point (herd of about 40). These mostly belong to Mr John Mackenzie and derive from stock owned by his father and possibly his grandfather. In 1951 a dun, hornless billy was introduced and as a result there are now some chocolate-coloured, hornless goats which are bigger-boned than the original, mostly dark grey, stock.

Scoraig (7)

Before World War II many goats, some feral, on the coast opposite Ullapool, between Ferry Inn and tip of Scoraig Peninsula; a few left in 1952 probably still survive including remnants of the herd (about 100 in 1937) running wild on Beinn Ghobhlach.

Slattadale (18)

On Forestry Commission plantation of Slattadale, near Tollie, goats formerly numerous; a few black billies and grey nannies still seen.

Strathconan (29)

A ghilly (John Macrae) early in this century ran 6 or 7 goats with his sheep between Balnault and the east march with Scatwell Forest at Curin; by the beginning of World War II there were still 5 but by 1945 only a nanny and a billy remained both of which subsequently died. North of Strathconan goats formerly seen on the forest of Strathbran probably wanderers from Torrachilty.

Strome (23)

Until 1930 large numbers of white goats roamed promontory between Loch Kishorn and Loch Carron, including Forestry Commission property of North Strome. By 1952 majority had been killed.

Tollie (15)

An apparently old-established herd (about 20) reported on Tollie near Poolewe.

Torrachilty (30)

Millais (190) mentions goats in the Strathpeffer district, probably referring to the goats on Torrachilty which have existed for at least 70 years. Until 1951 about 26 goats on the hill but Forestry Commission destroyed the majority and they were extinct by 1959.

Millais (1914) also mentions stalking a white billy in the mountains near Garve, possibly at Wyvis, but there are no goats there today.

Roxburghshire

Wauchope (134)
   A herd, fluctuating between 10 and 30 in recent years, on
   Forestry Commission property of Wauchope (Lethem)
   south of Jedburgh. Further south wanderers occur in
   the Cheviots and a goat is seen occasionally near the
   Forestry Commission property at New Castleton. In
   1958 Mr H. Tegner saw 2 goats just south of Carter's Bar
   on the Roxburgh-Northumberland border and believed
   a herd of 7-8 in the area.

Selkirkshire

St Mary's Loch (131)
   Goats around St Mary's Loch and up the Megget Water
   glen have been extinct for many years, but there are some
   around Birkhill on the Dumfries border. Occasionally
   they come on Herman Law, Bell Craig and Bodesbeck
   Law which lie west of Ettrick Water. There have been
   goats in the Ettrick-Eskdale district for over 150 years.

Stirlingshire

Rowardennan and *Ben Lomond* (103)
   Goats associated with Ben Lomond for over 600 years,
   since the time of Robert Bruce or earlier, though present
   stock appears to be of more recent origin. Their headquarters are the steep eastern scarp from Rowardennan
   to about the county boundary beyond Inversnaid, the
   main concentration being near Ptarmigan; also often seen
   on Craig Rostan. Recent estimates around 140-200—
   about double the 1952 estimate; formerly over 300 but
   numbers reduced by Forestry Commission. Early this

century only a few goats in the area, present stock
descended from domestic goats originally kept at
Inversnaid and Frenich. During World War I there was
an army camp near Loch Ard whose milking goats were
liberated when the camp was disbanded, and these joined
the herds on Ben Venue and Loch Lomondside.

Sutherland

Ben More Assynt (2)
Between the two wars there were goats on the lower
slopes of the east side of Ben More Assynt, their present
status is uncertain. About 1930 a billy reported almost
opposite Overscaig on Loch Shin.

Kinlochbervie (3)
Small herd at Rimichie, on the south shore of Loch
Inchard facing Kinlochbervie; a few on the rocks above
Laxford Bay (5 miles south of Kinlochbervie); present
status uncertain.

Loch Inver (4)
A few, including a fine billy, reported (1952) from an
island (presumably Soyea Isle).

The Mound (1)
Goats on the Mound Rock, Aberscross at the head of
Loch Fleet (near Rogart) for many years but present
stock (about 15 in 1963) not truly feral. Herd at Aberscross (mentioned in 17th century) died out in 1947 except
an old billy who eventually died after falling from
Morvich Rock about a mile from the Mound.

SCOTLAND: WEST COAST ISLANDS
(*map on page* 148)

Argyllshire

Cara (97)

Goats present certainly from 18th century; present stock white and typically Saanen, dark animals rare and have been shot in recent years. About 1930 goats from the island of Jura introduced; normal size of herd about 20.

Coll (68)

Goats present in 18th and 19th centuries but now extinct.

Colonsay (87)

A herd (18 in 1952) in the north around Balnahard and another (16 in 1952) in the centre around Machrins soon increased to over 100. One of the two domestic Saanen nannies kept by the minister on Colonsay mated with a wild billy from the Balnahard herd and the offspring (a billy) was liberated in 1949.

Gunna (67)

Goats reported in 1934 from small island of Gunna (between Coll and Tiree), probably extinct.

Islay

Gortantaoid (91); *Staoisha* (95)

Altogether about 70 (1952) in these areas, split into small herds. Gortantaoid stock introduced about 1914 from Mull and Jura; mostly white. Fresh blood possibly introduced occasionally from domestic goats kept at lighthouse at Rudh a Mhail point.

Kinnabus (93)

Herd at Kinnabus on the Mull of Oa (southern promontory of Islay) probably the oldest of the Islay herds; almost 300 reported in the 1960s.

Smaull and *Braigo* (92)
Herds on the west coast of Islay centre around Smaull and Braigo; these goats (40-60 in 1952) were introduced from Kinnabus possibly about the end of 18th century. About 1954 some from Smaull were taken to Lossit Farm on the Rhinns of Islay; 10 years later there were about 20 white animals with horns there.

Jura
Ardfin and *Craighouse* (90)
In south east Jura about 50 around Craighouse, coming into woods around Ardmenish Farm during the winter; not very wild and mostly white or black and white with a few light brown—the last the result of the introduction of a pair of goats from N. Ireland by a sailor on the Irish steamer run. Early this century some goats were introduced from Laggan on the Lochbuie estate, and about 1927 some billies were sent to Mull in exchange for more billies from Lochbuie. West of Craighouse about 20 frequented (1959) the Forestry Commission ground but were driven off when area was fenced.

Corpach and *Ardlussa* (89)
In north Jura about 200 mostly dark brown goats on coast between Loch Tarbert and Corpach Bay; further north on the Ardlussa ground perhaps 100; these move around in small herds of about 20.

Lismore (76)
Goats once plentiful but scarce by 1845; single billy reported in 1952.

Little Colonsay (81b)

Goats introduced about 1922, numbered 10-12 in 1937; still present in 1963.

Mull

Ardmeanach (83)

Goats on about 4 miles of sea cliff on the Ardmeanach Peninsula, including estate of the Burgh (National Trust for Scotland), Tavool and possibly part of Balmeanach.

Ben More (79)

A billy and 2 nannies (all white) introduced from Holy Island to the Ben More estate in 1947; occasional stragglers wander in from Lochbuie.

Carsaig (84)

Between 150-200 on coast between Carsaig Bay and Uisken; small herd (about 25) often seen near Bunessan, just north of Uisken. Occasional stragglers on Beinn-an-Aoinidh.

Lochbuie (78)

About 300-400 mostly on west and south shore of Loch Buie; some billies introduced from near Liverpool shortly before World War I. In 1920s billies exchanged with some from Jura and some years later some Lochbuie billies sent to Rhum.

Oronsay (88)

Herd of about 30, dark brown to black goats; fresh blood of unknown origin introduced 1947.

Seil (85a)

Herd of 40-50 decimated by poachers during World War II, now extinct.

Shuna (74)

By 1964 goats reduced to 2 nannies; owner attempting to re-establish herd

Texa (94)

Goats on Texa, a small uninhabited island lying off the south-east coast of Islay (about 12 in 1964) descended from domestic stock kept by former inhabitants and are mostly white.

Ulva (81a)

In the 1930s herd of about 25 marooned themselves on tidal islet in the Sound of Loch Tuath and about 20 drowned; similar accident in 1940s exterminated the herd.

Davaar (99); *Easdale* (85b); *Eorsa* (80); *Gigha* (96); *Scarba* (86); *Staffa* (82); *Tiree* (66); *Kerrera*

Goats extinct on all islands, except possibly Davaar and Kerrera, where introduction recent.

Ayrshire

Ailsa Craig (119)

Goats present certainly from 18th century; in this century a herd of about 60 reduced and finally exterminated in 1925 leaving only domestic stock which recently numbered about 12.

Buteshire

Bute (114)

Herd in south near Garroch Head introduced early this century and killed off shortly before World War II. Herd in north of more ancient origin; about 50 in 1949 reduced to under 12 by 1952.

Arran (115)

No wild goats on Arran for at least 100 years.

Holy Island (116)

Up to 1874 the tenant of Holy Island Farm had some goats of Irish origin, some hornless. In 1874 grey goats from Mull, Campbeltown and Rothesay were introduced and in 1906 a white billy from Mrs Moreton Macdonald of Largie; all coloured animals were then removed and an all-white herd built up. Herd of 40 (1950), 39 (1969).

Inverness-shire

Eigg (59)

Before World War I the Animal Breeding Department of the University of Edinburgh instigated an experimental study by introducing selected animals to this island which then had no goats. Four nannies (1 old English Type, 1 white Saanen, 1 pedigree British Alpine and 1 crossbred with some Anglo-Nubian blood) together with 2 other nannies and a billy (from Holy Island in Lamlash Bay, Arran) were introduced. The herd increased (about 40 in the 1930s), diminished during the war and died out in the winter of 1946-7.

Harris (65)

Goats present at end of last century, possibly a few remain near Loch Resort. Early this century about 15 near Taran (south side of Loch Resort), supposed to have been left by a shepherd who emigrated to Lewis, and now extinct.

Rhum (60)

Domestic goats noted by Pennant (1772). Before World War II about 110 on the western cliffs, reduced by 1947 to 80; mostly dark coloured, many billies almost black with very long hair. Two billies introduced (about 1926) from Inverinate, Kyle of Lochalsh and 2 more (1928) from Lochbuie, Mull. Herd of 8-10 white goats on Traill-mheall and Ais-mheall (Ashval) disappeared about 1934.

Scalpay (64)
Goats present towards the end of the last century now extinct. British Museum has heads of 2 white goats from Scalpay, dated 1895.

Skye
Suisnish (62); *Kylerhea* (61); *Ben Meabost* (63)
Goats on headland between Rudha Suisnish and Boreraig possibly descended from domestic stock left by 17th century crofters; before 1930 about 40, reduced to 11 by 1949 and to 3 billies by 1952. East of Boreraig goats formerly frequented shore between Kinloch Lodge and Kylerhea; now extinct. Also now apparently extinct is herd on Ben Meabost. Some goats reported on the Cuillins in 1950.

Ross & Cromarty

Summer Islands (6)
In 1936 Dr Fraser Darling took 2 newly kidded goats to Priest Island (4 miles west of Cailleach Head) and these soon went feral. In 1937 a few goats were reported from Horse Island.

IRELAND: EIRE
(map on page 156)

Wild goats in Ireland were little molested before World War II, except by sportsmen seeking a good head or local inhabitants collecting a kid for the pot. During the war years, however, many goats were killed for use as animal food in Phoenix Park Zoo, Dublin, or exporting to England for human consumption. As a result many herds were greatly reduced in numbers and some wiped out.

Co Clare
Burren Hills (208)

Before World War II probably hundreds of goats ranging over some 20 square miles of limestone crags and scrub; many killed during the war and only a few remain.

Co Donegal
Gweedore and *Derryveagh Mountains* (188)

In Donegal a few frequent the Glenveagh district, especially around Creenarg. During last century there were goats in the Gweedore and Derryveagh Mountains.

Co Dublin
Howth and *Baily* (192); *The Scalp* (193)

Goats on headland opposite Howth and Baily Lighthouse not truly feral. There were some goats in 1940s on the Scalp (on the Dublin-Wicklow boundary).

Co Galway
Costelloe (209)

In recent years about 100 lived on the lake shores between Leam and Costelloe, often swimming out to the islands when disturbed.

Joyce's Country (213)

Small herds occur in the mountainous region of northern Connemara known as Joyce's Country, probably also along the coast and on some islands; goats apparently now extinct on Aran Island, Galway Bay.

Lettercraffroe (210); *Moycullen* (211)
 A few may still remain around Lough Lettercraffroe but
 to the east the Moycullen district has been afforested and
 goats exterminated.

Twelve Pins Mountains (212)
 A herd of about 50 near Kylemore Lake.

Co Kerry
Innisteeraght Island (206); *Innisvickillane Island* (205);
Skellig Rocks (204)
 A pair introduced (about 1880) to Innisteeraght bred
 prolifically until by 1895 they had become a nuisance.
 Many were then killed and about 20 taken to Innisvickillane where they increased to some 180 by 1911,
 but have now declined and may be already extinct. A few
 exist on the barren Skellig Rocks, off St Finans Bay.

Mt Eagle and *Brandon* (207)
 Wild goats on the Dingle Peninsula for over 90 years. On
 Mt Eagle and the Ballydavid side of Mt Brandon a herd
 of 150 mostly brown goats has become established during
 past 60 years. A herd of white goats is (1950) also
 reported from the Peninsula.

Torc and *Mangerton Mountains* (203)
 Small herd has its headquarters on Torc Mountain and
 ranges over the adjoining area.

Co Mayo
Achil Island (190)
 Goats have lived here for over 200 years; earliest known
 strain blue-grey with immense sweeping horns in males,
 survived from the 18th century until early in the 20th.
 At the end of the 19th century a herd of long-haired

white goats frequented Corrymore and Croaghaun but present stock (about 100 in 1950, less now) is of recent origin, deriving from animals introduced around 1900.

Curraun Peninsula (191)

Goats on the mainland opposite Achil Island have existed for over a century; 15 seen in 1957.

Co Sligo

Mountains of Sligo (189)

A few reported recently.

Co Tipperary

Slieve-na-Mon (202)

Recently about 6 were reported here.

Co Waterford

Comeragh Mountains (201)

A few still frequent the Comeragh Mountains.

Co Wicklow

Bray Head (195); *Great Sugar Loaf* (196)

On coast at Bray head a large herd of not truly feral goats live principally about the railway cuttings. About 2 miles inland there used to be a herd on the Great Sugar Loaf mountain—R. F. Scharff presented a pair of horns from the supposed leader of this herd to the Dublin Museum.

Hollywood Glen (199); *Glenmalur* (200)

Before World War II there were goats on the mountains south of Glenmalur but there are none today. A large herd existed on cliffs of Hollywood Glen, near Donard but area has been afforested and goats probably eliminated.

Luggala (197); *Glencree* (194); *Glendalough* (198)
About 20 in vicinity of Luggala lake; to the north a few remain at Glencree (west of Powerscourt), to the south about 60 on cliffs overlooking Glendalough Lake.

IRELAND: NORTHERN IRELAND

Goats occur in all the counties with the possible exception of Londonderry and Tyrone.

Co Antrim
Fair Head (180)
A few still live on Fair Head; 5 nannies and 1 billy seen in 1959. Dead billy found on Ballycastle beach in 1954 believed to have come from Fair Head.

Garron Point (181)
Some 15 miles south of Fair Head a herd (about 10) lives on Garron Point. South of Garron Point a few goats exist near Glenarm.

Glenarm (182)
In 1959, 21 were reported near Dunemencock.

Knockagh (183)
Goats present during the last century now extinct.

Rathlin Island (179)
About 20 on cliffs usually near the West Light possibly

145

derive from those liberated by owner of the Island about 1760.

Co Armagh

Camlough Mountain (185)
Herd (about 20) frequents the Camlough Mountain in southern Armagh.

Slieve Gullion (186)
Several herds in the area; in 1959 a census by District Forestry Officer gave the following result: Killeavy and Ballintemple 16; Shanroe, Co Loughside 23; Carrick-broad, Co Loughside 20; Slieve Gullion 9.

Co Down

Mourne Mountain (184)
Mourne Mountains famous for goats in the 18th century; a few recent reports of individual goats have not been substantiated.

Co Fermanagh

Lough Erne and *Lough Navar* (187)
Small numbers reported on some uninhabited islands in the lough; a few on the higher cliffs in the Lough Navar forest region.

Goat Distribution Maps

Map 1—Scotland

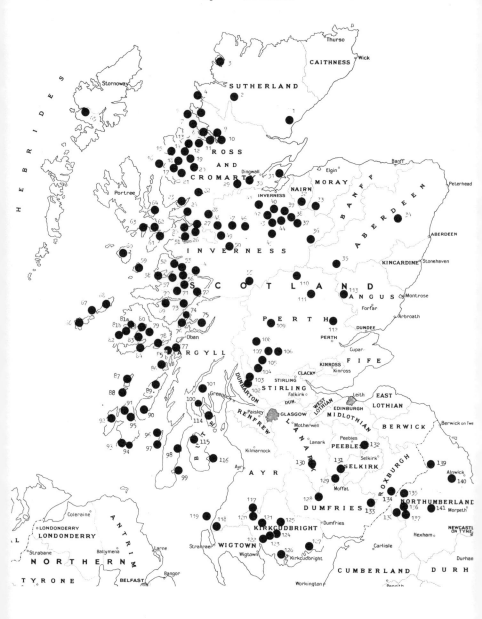

Map 1—Scotland

SUTHERLAND
1 The Mound
2 Ben More Assynt
3 Kinlochbervie
4 Loch Inver

ROSS & CROMARTY
5 Coigach
6 Summer Isles
7 Scoraig
8 Badrallach
9 Leckmelm
10 Lael
11 Dundonnell Hill
12 Anteallach
13 Gruinard
14 Fisherfield
15 Tollie
16 Red Point
17 Flowerdale & Shieldaig
18 Slattadale
19 Letterewe
20 Kinlochewe
21 Beinn Eighe
22 Achnashellach
23 Strome
24 Inverinate and Dorisduan
25 Ratagan
26 Glenshiel & Clunie
27 Kintail
28 Glomach
29 Strathconan
30 Torrachilty
31 Black Isle

NAIRNSHIRE
32 Findhorn

MORAYSHIRE
33 Lochindorb

ABERDEENSHIRE
34 Bennachie
35 Glencallater

INVERNESS-SHIRE
(including islands)
36 Rothiemurchus
37 Kinveachy
38 Slochd & Corrybrough
39 Upper Findhorn
40 Farr
41 Dochfour
42 Flichity (Brin)
43 Glenmazeran
44 Dalmigavie
45 Coignafearn
46 Port Clair

47 Glenmoriston
48 Affric
49 Ceannacroc
50 Invergarry
51 Loch Hourn
52 North Morar
53 Meoble
54 Achdalieu
55 Corrour
56 Loch Shiel
57 Inverailort
58 Eilean nan Gobhar
59 Isle of Eigg
60 Isle of Rhum
61 Kylerhea, Isle of Skye
62 Rudha Suisnish, Isle of Skye
63 Ben Meabost, Isle of Skye
64 Isle of Scalpay
65 Loch Resort, Isle of Harris

ARGYLLSHIRE (including islands)
66 Isle of Tiree
67 Isle of Gunna
68 Isle of Coll
69 Glencrippesdale
70 Glen Hurich
71 Conaglen
72 Ardgour
73 Kilmalieu
74 Isle of Shuna
75 Benderloch
76 Isle of Lismore
77 Ardmaddy
78 Lochbuie, Isle of Mull
79 Ben More, Isle of Mull
80 Isle of Eorsa
81a Isle of Ulva
81b Isle of Little Colonsay
82 Isle of Staffa
83 Ardmeanach, Isle of Mull
84 Carsaig, Isle of Mull
85a Isle of Seil
85b Isle of Easdale
86 Isle of Scarba
87 Isle of Colonsay
88 Isle of Oronsay
89 Corpach & Ardlussa, Isle of Jura
90 Ardfin & Craighouse, Isle of Jura
91 Gortantoid, Isle of Islay
92 Smaull & Braigo, Isle of Islay
93 Kinnabus, Isle of Islay
94 Isle of Texa (Islay)
95 Staoisha, Isle of Islay
96 Isle of Gigha
97 Isle of Cara
98 Carradale, Mull of Kintyre
99 Isle of Davaar
100 Tighnabruaich
101 Glendaruel

149

Map 1—Scotland (*continued*)

BUTESHIRE
114 Isle of Bute
115 Isle of Arran
116 Holy Island

DUNBARTONSHIRE
102 Inchlonaig Island

STIRLINGSHIRE
103 Rowardennan and Ben Lomond
PERTHSHIRE
104 Loch Ard
105 Ben Venue
106 Glenartney
107 Ardnandave
108 Braes of Balquhidder
109 Meggernie
110 Atholl
111 Schiehallion
112 Rohallion

ANGUS
113 Glenisla

PEEBLES-SHIRE
132 Peel

SELKIRKSHIRE
131 St. Mary's Loch & District

LANARKSHIRE
130 Evanfoot

AYRSHIRE
117 Kirrieroch Hill
118 Glen App
119 Ailsa Craig

KIRKCUDBRIGHTSHIRE
120 Merrick Hill
121 Glen Trool
122 Kirroughtree
123 Cairnsmore
124 Craignelder
125 Cairn Edward
126 Cairnharrow
127 Dalbeattie

DUMFRIES-SHIRE
128 Queensbury Hill
129 Grey Mare's Tail and
    Craigieburn, inc. Moffat Water
133 Langholm, inc. Megget Water

ROXBURGHSHIRE
134 Wauchope

ADJACENT TO THE
ENGLISH BORDER
135 Kielder
136 Plashetts
137 Simonburn and Wickhope
138 Christianbury Crag
139 Cheviot and College Valley
140 Thrunton Crag
141 Callerhues

Map 2—England (North, East and Midlands)

135 Kielder, Northumberland
136 Plashetts, Northumberland
137 Wickhope and Simonburn, Northumberland
138 Christianbury Crag, Cumberland
139 Cheviot and College Valley, Northumberland
140 Thrunton Crag, Northumberland
141 Callerhues, Northumberland
142 Lowther Park, Westmorland
143 Martindale, Westmorland
144 Coniston, Lancashire
145 Levens Park, Kendal, Westmorland
146 Kilnsey Crag, Yorkshire
147 Otley Chevin, Yorkshire
167 Tatton Park, Cheshire
168 Calke Abbey, Derbyshire
169 Ingestre Park, Staffordshire
170 Bagot's Park and Blithfield, Staffordshire
171 Woburn, Bedfordshire
172 Easton Park, Suffolk

NOTE :—Feral goats exist on the Isle of Man (148, 149)—see Map 5
       For south-east England and south-west England, see maps 3 and 4
       respectively

Map 2—England (North, Eastern Midlands)

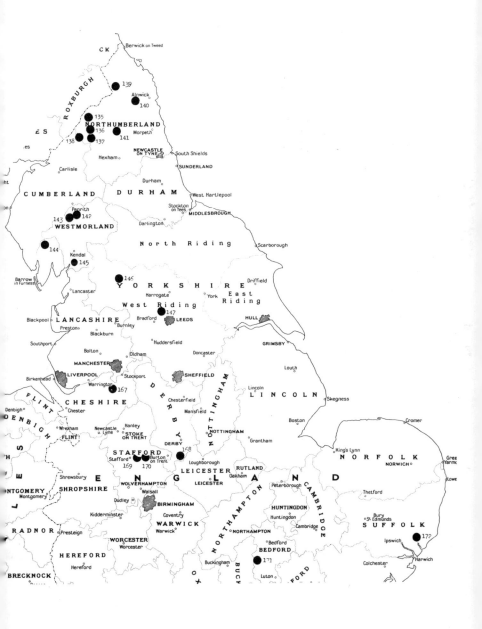

Map 3—England (south-east)

ster
STER
Oxford °
Aylesbury °
Hertford °
ESSEX
Chelmsford °
FORD
'INGHAM
HERT
Swindon °
BERKSHIRE
Windsor °
MIDDLESEX
LONDON
Southend-on-Sea °
Margate
Newbury °
READING
173
Kingston
Croydon °
Chatham
Devizes °
Maidstone °
LTSHIRE
Basingstoke °
SURREY
KENT
Guildford °
Dover
Salisbury °
Winchester °
Tunbridge Wells °
HAMPSHIRE
SUSSEX
SOUTHAMPTON
Lewes °
T
Chichester °
174
BRIGHTON
Eastbourne °
PORTSMOUTH
ISLE OF WIGHT
Bournemouth

Map 4—England (south-west)

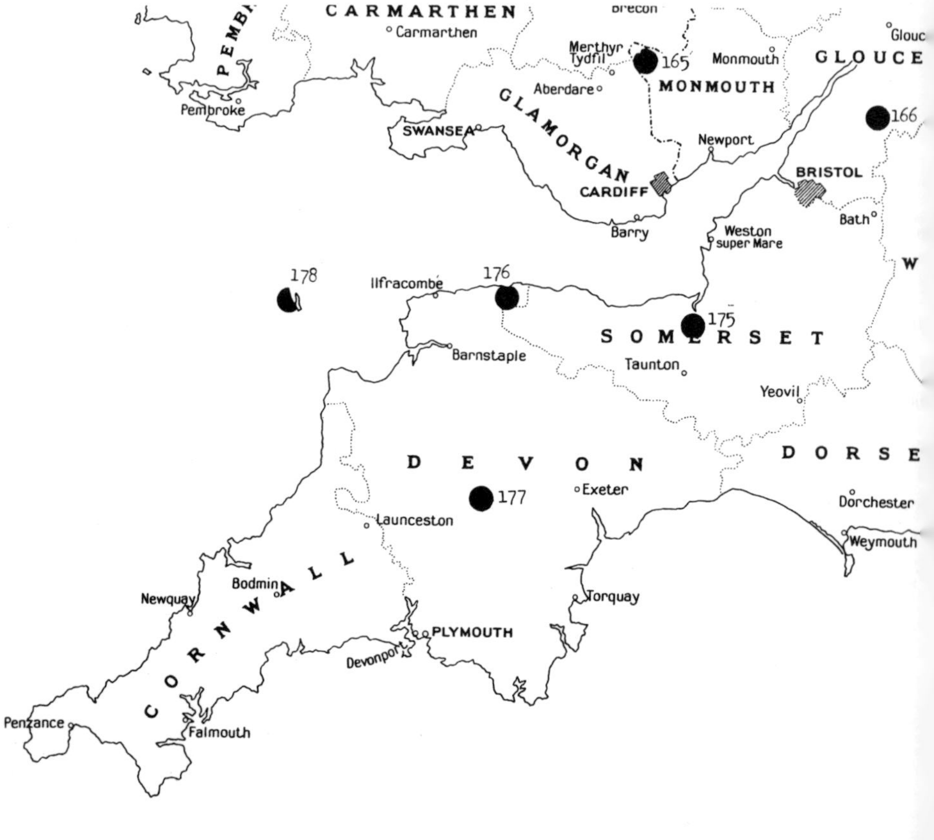

CARMARTHEN
Carmarthen °
Brecon
Glouc °
PEMB
Merthyr Tydfil °
165
Monmouth °
GLOUCE
Aberdare °
MONMOUTH
SWANSEA
GLAMORGAN
Newport
166
Pembroke
CARDIFF
BRISTOL
Barry °
Bath °
Weston super Mare
W
178
Ilfracombe °
176
SOMERSET
175
Barnstaple °
Taunton °
Yeovil °
DEVON
DORSE
Dorchester °
177
Exeter °
Weymouth
Launceston °
Torquay °
Bodmin °
Newquay °
PLYMOUTH
CORNWALL
Devonport
Penzance °
Falmouth °

Map 3—England (south-east)

173 Windsor Park, Berkshire
174 Arundel Park, Sussex

Map 4—England (south-west)

165 Tredegar Park, Monmouthshire
166 Kingswood, Gloucestershire
175 Athelney, Somerset
176 Lynton Rocks, Devon
177 Rough Tor, Devon
178 Lundy, Devon

Map 5—*Wales and Isle of Man*

Map 5—Localities in Wales and the Isle of Man where Feral Goats exist or formerly existed

148 Isle of Man
149 Calf of Man
150 Great Ormes Head, Caernarvonshire
151 Amlwch, Anglesey
152 Glyders and Tryfan, Caernarvonshire
153 Snowdonia, Caernarvonshire
154 Mynydd Mawr, Caernarvonshire
155 Moel Siabod, Caernarvonshire
156 The Rivals, Caernarvonshire
157 Tremadoc Rock, Caernarvonshire
158 Moelwyn, Merionethshire
159 Craig-y-Benglog, Merionethshire
160 Rhinogs, Merionethshire
161 Cader Idris, Merionethshire
162 (Fishguard, Pembrokeshire
      (Dinas, near Fishguard, Pembrokeshire
163 Prescelly, Pembrokeshire
164 Craig Pwll Du, Radnorshire

Map 6—Ireland

Map 6—Localities in Ireland where Feral Goats exist or formerly existed

NORTHERN IRELAND

179 Rathlin Island, Co. Antrim
180 Fair Head, Co. Antrim
181 Garron Point, Co. Antrim
182 Glenarm, Co. Antrim
183 Knockagh, Co. Antrim
184 Mourne Mountain, Co. Down
185 Camlough Mountain, Co. Armagh
186 Slieve Gullion, Co. Armagh
187 Lough Erne and Lough Navar, Co. Fermanagh

EIRE

188 Gweedore & Derryveagh Mountains (inc. Glenveigh), Co. Donegal
189 Mountains of Co. Sligo
190 Achil Island, Co. Mayo
191 Curraun, Co. Mayo
192 Howth & Baily, Co. Dublin
193 The Scalp, Co. Dublin
194 Glencree, Co. Wicklow
195 Bray Head, Co. Wicklow
196 Great Sugar Loaf, Co. Wicklow
197 Luggala, Co. Wicklow
198 Glendalough, Co. Wicklow
199 Hollywood Glen, Co. Wicklow
200 Glenmalur, Co. Wicklow
201 Comeragh, Co. Waterford
202 Slieve-na-Mon, Co. Tipperary
203 Torc and Mangerton Mountain, Co. Kerry
204 Skellig Rocks, Co. Kerry
205 Innisvickillane Island (Blasket Island Group), Co. Kerry
206 Innisteeraght Island, Co. Kerry
207 Mount Eagle and Brandon, Dingle Peninsula, Co. Kerry
208 Burren Hills, Co. Clare
209 Costelloe, Co. Galway
210 Lettercraffoe, Co. Galway
211 Moycullen, Co. Galway
212 Twelve Pins Mountain, Co. Galway
213 Joyce's Country, Co. Galway

APPENDIX—RECORD GOAT HEADS

(i)   Wild (feral) Goat Heads—Scotland.

(ii)   Wild (feral) Goat Heads—England.

(iii)   Wild (feral) Goat Heads—Wales.

(iv)   Wild (feral) Goat Heads—Ireland.

(v)   Domesticated Goat Heads—World.

In these five tables, the source of information for the measurement of individual heads has been as follows:—

1. Baroness Burton.

2. *The Field.*

3. R. G. Fletcher.

4. T. Russell Goddard.

5. J. Macdonald.

6. Farquhar Macgregor.

7. J. G. Millais.

8. Owner.

9. Margery I. Platt.

10. T. Pennant.

11. A. D. Buchanan Smith.

12. Rowland Ward, Ltd.

13. H. Boyd Watt.

14. G. Kenneth Whitehead.

APPENDIX (i)
WILD (FERAL) GOAT HEADS: SCOTLAND

Date	Locality	Length inches	Circ at base inches	Tip to tip inches	Remarks	
Circa 1931	Isle of Bute	44¾	8¾	?	Found dead	(11)
?	Isle of Bute	39½	7³⁄₁₆	35¼	Owned by Lord Wavertree	(14)
Pre 1903	"Scotland"	37½	7¼	32¾	Owned by Walter Jones	(12)
1899	Meoble, Inverness	37	7½	35	Shot by D. Barry: 9 yr old goat	(12)
July 1958	Rowardennan, Stirlingshire	36⅜	7⅞	39	Shot by Mr Robertson on Kintail and later found dead on Inverinate	(14)
Circa 1935	Kintail, Ross	36	?	?		(6)
Pre 1906	Ross-shire	36	?	?		(7)
Pre 1903	"Scotland"	34¼	7½	38½	Owned by Duke of Bedford	(12)
Pre 1903	"Scotland"	34	8½	32½	Owned by British Museum	(12)
Pre 1922	Achray, Perthshire	34	8	27	Owned by Major B. M. Edwards	(12)
1891	Near Garve, Ross	34	?	?	Shot by J. G. Millais	(7)
1956	Ben Lomond, Stirling	33½	7⅝	35½	Shot by D. Barry, Jr: 8 yr old goat (approx)	(14)
Pre 1914	Meoble, Inverness	33½	8	26½	Owned by Hon O. C. Molyneux	(12)
Pre 1914	Cairnsmore, Kirkcudbright	33	7½	39	Owned by Major Champion	(14)
Pre 1906	Sutherland	33	?	?		(7)
Sept 1951	Rowardennan, Stirlingshire	32¾	7	33	Shot by D. Barry: 8 yr old goat (approx)	(14)

APPENDIX (i)
WILD (FERAL) GOAT HEADS: SCOTLAND (cont.)

Date	Locality	Length inches	Circ at base inches	Tip to tip inches	Remarks	
1944	Inversnaid, Stirlingshire	$31\frac{1}{2}$	$8\frac{5}{8}$	31	Shot by W. Joynson: 8 yr old goat (approx)	(14)
1947	Rowardennan, Stirlingshire	31	$8\frac{1}{4}$	$28\frac{3}{4}$	Shot by J. W. Barry: 8 yr old goat (approx)	(14)
?	Dochfour, Inverness	31	?	$36\frac{1}{4}$	Bigger goats have been shot on this ground	(1)
1905	Kintail, Ross	31	9	?	Hanging in Kintail Lodge	(6)
1928	Carsaig, Isle of Mull, Argyll	31	?	?	Shot by V. Hardwick	(8)
1899	Meoble, Inverness	$30\frac{1}{2}$	$7\frac{1}{2}$	35	Owned by Walter Jones	(12)
1944	Isle of Rhum, Inverness	$30\frac{1}{2}$	8	$28\frac{1}{2}$	Shot by G. Kenneth Whitehead	(14)
Circa 1920	"Scotland"	$30\frac{1}{2}$	?	?	Large white billy goat	(2)
?	Ardlussa, Isle of Jura, Argyll	29	?	?	There are better on the forest	(3)
1953	Letterewe, Ross	$28\frac{3}{4}$	?	?	Shot by M. W. Whitbread	(8)
1937	Ben Venue, Perth	$28\frac{1}{4}$	$7\frac{9}{16}$	29	Shot by W. Joynson: 7 yr old goat	(14)
Circa 1914	Barrs, Loch Etive, Argyll	28	?	?	Shot by G. Holmes	(8)
1955	Cairnsmore, Kirkcudbright	$26\frac{1}{2}$	$7\frac{1}{2}$	$21\frac{1}{2}$	Shot by G. Kenneth Whitehead	(14)
?	Isle of Jura, Argyll	$24\frac{3}{8}$	?	$27\frac{1}{16}$	In Royal Scottish Museum	(9)
?	Langholm, Dumfries	?	?	48		(13)

APPENDIX (ii)
WILD (FERAL) GOAT HEADS: ENGLAND

Date	Locality	Length inches	Circ at base inches	Tip to tip inches	Remarks	
?	Bagot's Park Wood, Staffs	$41\frac{1}{4}$	8	46	Aged about 10 years	(14)
?	Bagot's Park Wood, Staffs	38	$6\frac{3}{4}$	$38\frac{1}{4}$	Aged about 9 years	(14)
1958	Bagot's Park Wood, Staffs	$35\frac{1}{2}$	$7\frac{1}{8}$	43	Shot by G. Kenneth Whitehead	(14)
?	Bagot's Park Wood, Staffs	35	$7\frac{15}{16}$	35	Aged about 8 years	(14)
?	Bagot's Park Wood, Staffs	34	$7\frac{7}{16}$	$22\frac{3}{4}$	Aged about 8 years	(14)
?	Bagot's Park Wood, Staffs	$31\frac{3}{4}$	$7\frac{7}{16}$	39	Aged about 7 years	(14)
?	Cheviot Hills, Whickhope, Northumberland	31	?	?		(4)
?	Cheviot Hills, Whickhope, Northumberland	29	?	35		(5)

APPENDIX (iii)
WILD (FERAL) GOAT HEADS: WALES

Date	Locality	Length inches	Circ at base inches	Tip to tip inches	Remarks	
18th C	Glamorgan	44	?	36		(10)
Pre 1793	?	39	?	38		(10)
Pre 1906	Fishguard, Pembroke	37	9	38¼	Shot by Lord David Kennedy	(7)
Pre 1906	Fishguard, Pembroke	34	8	13		(7)
July 1959	Tryfan, Caernarvonshire	26	7½	22½	Shot by J. Henshaw	(8)
Pre 1899	South Wales	25½	7¾	4	Owner, Major G. Palmer	(12)

APPENDIX (iv)
WILD (FERAL) GOAT HEADS: IRELAND

Date	Locality	Length inches	Circ at base inches	Tip to tip inches	Remarks	
Circa 1953	Burren Hills, Co Clare	39¾	9	36	Found dead: owner G. Ross	(8)
Circa 1950	Glencree, Co Wicklow	33½	7¼	34½	Shot by L. F. O'Carroll	(8)
Circa 1948	Achil Island, Co Mayo	33	?	13	Owner, H. G. M. McDowell	(8)
?	Slieve Snacht, Co Donegal	27	?	?	Shot D. J. W. Edwardes	(8)
1957	Achil Island, Co Mayo	24½	7⅜	22½	Shot by G. Kenneth Whitehead	(14)
Circa 1880	Achil Island, Co Mayo	21	7½	15	Shot by E. G. Weldon	(8)
Circa 1955	Fair Head, Co Antrim	21	8	23	Found dead by Major A. E. Green	(8)

APPENDIX (v)
DOMESTICATED GOAT HEADS: WORLD

Date	Locality	Length inches	Circ at base inches	Tip to tip inches	Remarks	
Pre 1892	Daghestan	52½	10½	40⅝	G. Kenneth Whitehead Collection	(12)
Pre 1899	Angora	44¼	6	29¼	G. Kenneth Whitehead Collection	(12)
Pre 1899	Daghestan	40½	9⅜	3¾	G. Kenneth Whitehead Collection	(12)
Pre 1928	?	39	7	49½	Owner, 2nd Welch Fusiliers	(12)
Pre 1903	?	35¾	7¾	38¼	Owner, B. de Bertodano	(12)
Pre 1896	?	34½	8	25	Owner, Rowland Ward, Ltd	(12)
Pre 1922	?	34¼	7½	41	Owner, A. Browne	(12)
1959	?	33¼	7⅜	37¾	Owner, Welch Regiment	(14)

BIBLIOGRAPHY

The following is a list of the books, periodicals, etc which have been consulted during the preparation of this book. In many of them reference to goats, either wild or domesticated, may amount to only a paragraph or two. Other references have also been noted in the text.

Aaron, Margaret. 'Wild Goats in Wales'. Letter to *The Times*, (7 September 1951)

Airey, A. F. 'Wild Goats on Ben Lomond'. Letter to *Country Life*, (9 January 1948)

Alon, A. *Natural History of the Land of the Bible* (1970)

Alston, Edward R. 'On a 4-Horned Chamois', *Proc Zool Soc Lond* (1879)

Anderson, R. P. R. 'Wild Goats in Scotland'. Letter to the *Oban Times*, (8 March 1952)

Anon. 'Wild Goats in Wales', *Country Life*, (2 March 1901)

Anon. 'Goats in Wales', *Nature, Lond*, (18 January 1941) 83

Anon. 'On the Rocks', *The Times*, (21 August 1951)

Anon. 'Wild Goat Caught in Wales', *The Times*, (4 September 1951)

Anon. 'Day in the Country', *Sunday Dispatch*, (18 January 1959)

Anon. 'Goat Killings on Duke's Land Stir Protest', *Sunday Express*, (13 February 1966)

Appleford, W. P. 'A Hybrid Kashmir Goat and Himalayan Snowcock'. Letter to *The Field*, (5 October 1929)

A.R. 'Mare and Goat', *Countryman*, (Spring 1946), 57

Asdell, S. A. 'The Origin of the Goat', *Better Goatkeeping*, (1947), 537-39; 557-61

Asdell, S. A. & Buchanan Smith, A. D. 'Inheritance of Colour, Beards & Tassels or Wattles', *British Goat Society Year Book*, (1927), 129-132

Bagot, Richard. *Casting of Nets* (1901)

Baillie-Grohman, W. A. *Sport in the Alps* (1896)

Baker, A. L. L. 'Wild Goats in Wales'. Letter to *The Times*, (20 August 1951)

Baker, A. L. L. 'Wild Goats in Wales'. Letter to *The Times*, (3 September 1951)

Baker, A. L. L. 'Wild Goats in Wales', *The Field*, (2 September 1954)

Baker, A. L. L. 'Wild Goats in Wales'. Letter to *Country Life*, (24 October 1957)

Bartholomew (Berthelet), *Bartholomeus de Proprietatibus Rerum* (1535). Quoted by H. W. Seager, (see below)

Batty, T. H. 'English Goat'. Letter in *British Goat Society Monthly Circular*, 12 no 2 (1919), 3

Bentley, E. E. 'Goats' Milk for Calves'. Letter to *The Field*, (18 November 1954)

Bewick, T. *A General History of Quadrupeds* (Newcastle-upon-Tyne 1790)

Bingley, W. *Memoirs of British Quadrupeds* (1809)

Bingley, W. *Animal Biography, or, Popular Zoology* (6th edn 1824)

Blakeney-Scott, F. E. 'Goats in Ireland', *British Goat Society Year Book*, (1927), 127-8

Boyd Watt, Hugh. 'On the Wild Goat in Scotland', *Anim Ecol*, 6 (1937), 15-22

Braine, A. *The History of Kingswood Forest* (1891), 59

British Goat Society. 'What is an English Goat?', *BGS Monthly Circular*, 12 no 1 (1919), 5-9

British Goat Society. 'British Wild Goats', *BGS Monthly Circular*, 12 no 8 (1919), 2-3

British Goat Society. 'A Treatise on Goats—Written One hundred and Twenty-five Years Ago', *BGS Monthly Circular*, 12 no 9 (1919)

British Goat Society. 'Goats in Switzerland', *BGS Monthly Circular*, 13 no 7 (1920), 1-5

British Goat Society. 'British Breeds of Goats', *BGS Monthly Circular*, 13 no 7 (1920), 6-10

British Goat Society. 'British Wild Goats', *BGS Monthly Circular*, 13 no 9 (1920), 3

British Goat Society. 'British Wild Goats', *BGS Monthly Circular*, 13 no 11 (1920), 5-7

British Goat Society. 'The Wild Goats of Achil', *BGS Monthly Circular*, 15 no 11 (1922), 210-12

British Goat Society. 'The Antient Wild Goats of Wales', *BGS Year Book*, (1923), 52-5

166

British Goat Society. 'Wild Goats', *BGS Monthly Journal,* 24 no 1 (1931), 11

British Goat Society. 'Wild Goats', *BGS Monthly Journal,* 26 no 11 (1933), 275

British Goat Society. 'The Age of the Goat shown by the Teeth', *BGS Year Book,* (1956), 57

Brown, Thomas. *Anecdotes of Quadrupeds* (Glasgow 1831), 561-5

Brydone, P. C. 'Pine Marten, Pole Cats and Wild Goats'. Letter to *Oban Times,* (21 March 1963)

Buchanan Smith, A. D. 'Goats for the Island of Eigg'. *British Goat Society Year Book,* (1927), 57-8

Buchanan Smith, A. D. ' "Wild" Goats in Scotland', *British Goat Society Year Book,* (1932), 120-5

Buffon, Count de. *Natural History, General and Particular.* Translation by William Smellie, 4 (1812), 292-307

Campbell, G. Askew. 'Wild Goats in Scotland'. Letter to *Oban Times,* (29 March 1952)

Carne, P. H. 'Wild Goat Country', *The Gamekeeper & Countryside,* (July 1965)

Caseby, John A. 'The Welsh Goat', *British Goat Society Year Book,* (1936), 120-3

C.D. 'Goat Not a Mascot'. Letter to *Daily Telegraph,* (9 June 1953)

Chapman, Abel. *The Borders and Beyond* (1924)

C.J. 'Wild Goats'. Letter to *The Field,* (13 November 1920)

C.J.S. 'Strange Companions'. Letter to *The Field,* (20 January 1955)

Colquhoun, John. *The Moor and The Loch* (2nd edn 1841), 86-8

Corballis, J. H. *Forty-five Years of Sport* (1891), 393

Courage, R. 'Goats in the Stable'. Letter to *The Field,* (7 April 1955)

Couser, E. 'Goat Herd in Ireland'. Letter to *Country Life,* (26 December 1963)

Cox, J. Charles. *The Royal Forests of England* (1905)

Cox, N. *The Gentleman's Recreation* (6th edn 1721)

Crook, Ian. 'Feral Goats of North Wales', *Animals,* 12 no 1 (1969), 13-15

Crook, Ian. 'Wild Goats or Film Extras'. Letter to *Animals*, (August 1969)

Cummings, L. 'Goats on Islay'. Letter to *Country Life*, (6 April 1929)

Dalimier, Paul. 'La Morphologie de la Chevre sous L'influence de la Domestication', *Bull Inst & Sci nat Belg*, 30 no 13 (1954), 1-12

Darling, F. Fraser. 'Habits of Wild Goats in Scotland'. Supplement to paper 'On the Wild Goat in Scotland', by H. Boyd Watt, (see above)

Darling, F. Fraser. *Natural History in the Highlands and Islands* (1947), 75

Davenport, F. R. 'Wild Goats on Rhum'. Letter to *The Field*, (7 June 1941)

Davie, T. M. 'Goat Superstitions'. Letter to *The Field*, (5 April 1952)

Dawson, Monica. 'Goats' Milk for Calves'. Letter to *The Field*, (11 November 1954)

De Vos, Anton, Manville, R. H. & van Gelder, R. G. 'Introduced Mammals and Their Influence on Native Biota', *Zoologica, NY*, 41 section 19 (1956)

Dodds, W. 'Goats of the Border'. Letter to *Country Life*, (9 May 1952)

Dorst, J. & Dandelot, P. *A Field Guide to the Larger Mammals of Africa* (1970)

Douglas, Norman. *Birds and Beasts of the Greek Anthology* (1928)

Dunkeld, M. D. 'The Meaning of a Saying'. Letter to *The Field*, (12 February 1959)

Edlin, H. L. *The Changing Wild Life of Britain* (1952), 57-8

Edlin, H. L. *Wild Life of Wood and Forest* (1960), 42-4

Edwardes, D. J. W. 'An Irish Wild Goat'. Letter to *The Field*, (10 May 1930)

Edwards, E. ' "Wild" Goats in Scotland'. Letter to *Country Life*, (21 March 1957)

Elliot, Henrietta. 'Goats' Milk for Calves'. Letter to *The Field*, (11 November 1954)

Evans, Rex. 'Goat-Sheep Hybrids'. Letter to *The Field,* (14 October 1954)

Fisher, J. *et al. The Red Book* (1969)

Fitter, R. S. R. *The Ark in our Midst,* (1959), 53-9

Forbes, A. R. *Gaelic Names of Beasts* (Mammalia). *Birds, Fishes, Insects, Reptiles,* etc (Edinburgh 1905)

Forrest, H. G. *The Vertebrate Fauna of North Wales* (1907)

Forrest, H. G. *A Handbook to the Vertebrate Fauna of North Wales* (1919)

Gillies, P. H. *Netherlorn Argyllshire & Its Neighbourhood* (1909), 95 & 104

Goddard, Ruth A. 'Wild Goats in Snowdonia'. Letter to *Country Life,* (20 February 1948)

Goldsmith, Oliver. *A History of the Earth and Animated Nature* (1822)

Gowan, L. 'Wild Goats'. Letter to *The Field,* (24 September 1959)

Greswell, W. II. P. *The Forest and Deer Parks of the County of Somerset* (Taunton 1905)

Gubernatis, Angelo de. *Zoological Mythology,* 1 (1872), 400-32

Hafez, E. S. (ed). *The Behaviour of Domestic Animals* (2nd edn 1969)

Hamerton, P. G. *Chapters on Animals* (4th edn 1883)

Harris, Mary Corbett. 'Goats of the Welsh Mountains'. Letter to *Country Life,* (19 December 1963)

Harrison, D. L. *The Mammals of Arabia,* 2 (1968)

Harvie-Brown, J. A. & Buckley, T. E. *A Vertebrate Fauna of the Outer Hebrides* (Edinburgh 1888), 42

Harvie-Brown, J. A. & Buckley, T. E. *A Vertebrate Fauna of Argyll and the Inner Hebrides,* (Edinburgh 1892), 45

Hayes, J. L. *The Angora Goat* (New York 1882)

Heaton, W. H. *Zoologist,* (August 1878)

Henderson, R. B. 'Wild Goats in Wales'. Letter to *The Times,* (30 August 1951)

Her Majesty's Stationery Office. *Loch Ard* (1951)

Higgins, Frank. 'Welsh Wild Goats'. Letter to *The Field,* (30 July 1953)

Hingston, R. W. G. *The Meaning of Animal Colour and Adornment* (1933)

Holmes, W. F. 'Wild Goats in Wales'. Letter to *The Times,* (27 August 1951)

Hook, Bryan. *Milch Goats and Their Management* (1896)

Hopkinson, S. R. 'Goats of the Welsh Mountains'. Letter to *Country Life,* (19 December 1963)

Hortus Sanitatis (c1500). Quoted by H. W. Seager, (see below)

Hulme, F. E. *Natural History, Lore and Legend* (1895)

Jarvie, J. K. 'Inverness-shire Goats'. Letter to *Country Life,* (16 January 1964)

Jeffery, H. E. *Breeds of Goats,* British Goat Society Pamphlet (N.D.)

Johns, W. E. 'Adders and Grass Snakes'. Letter to *The Field,* (10 November 1951)

Johnston, H. *British Mammals* (1903), 351

Joicey, M. E. 'Wild Goats of Britain'. Letter to *Country Life,* (24 October 1947)

Kennion, R. L. 'Hybrids between Domestic Goat and Ibex'. Letter to *The Field,* (16 November 1929)

Knapp, C. A. 'Horses and Goats'. Letter to *The Field,* (29 December 1955)

'Lichen Grey'. 'Goat Superstitions', *Country Life,* (21 October 1905)

Lockley, R. M. *The Island Farmers* (1946)

Loder, John de Vere. *Colonsay & Oronsay in the Isles of Argyll, their History, Flora, Fauna and Topography* (Edinburgh 1935), 185

Low, David. *On the Domesticated Animals of the British Isles* (1845)

Lupton. *A Thousand Notable Things of Sundry Sortes* (1627). Quoted by H. W. Seager, (see below)

Lydekker, R. *Wild Oxen, Sheep & Goats of All Lands* (1898)

Lydekker, R. *Mostly Mammals* (1903)

MacDonald, Charles. *Moidart,* (Oban 1889)

MacDonald, May A. 'Goats' Milk for Calves'. Letter to *The Field,* (18 November 1954)

Mace, Herbert. 'The English Goat', *The Field*, (15 January 1925)

MacIntyre, D. 'Wild Goats in Scotland'. Letter to the *Oban Times*, (22 March 1952)

Mackenzie, David. *Goat Husbandry* (3rd edn 1971)

Mackenzie, D. A. *Myths of Babylonia & Assyria* (N.D.)

Mackenzie, O. H. *A Hundred Years in the Highlands* (1922)

MacNally, L. Scottish Scene. The Habits of Pinemartens, Wild Goats and Badgers, *Shooting Times & Country Magazine* (18 February 1965)

Magnus, Albertus. *Virtue of Animals* (1553). Quoted by H. W. Seager, (see below)

Manwood, John. *A Treatise of the Laws of the Forest* (3rd edn 1665)

Maplet, John. *A Greene Forest* (1567, reprinted 1930), 148

Martin, M. *A Description of the Western Islands of Scotland* (1703)

Matthews, L. Harrison. *British Mammals* (1952)

Maxwell, W. H. *Wild Sports of the West* (c1830)

Maxwell, W. H. *The Field Book* (1833), 221

McConnochie, A. I. *The Deer & Deer Forests of Scotland, Historical, Descriptive, Sporting* (1923)

McTaggart, H. S. 'Feral Goats, Holy Island, Arran', UFAW Annual Report, (1968-9) 22-4

Metge, R. A. 'Goats and Cows'. Letter to *The Field*, (3 May 1952)

Middleton, Lord. 'Goats in Stables'. Letter to *The Field*, (29 April 1955)

Millais, J. G. *The Mammals of Great Britain & Ireland*, 3 (1906), 213-14

Millais, J. G. *The Big Game of Africa and Europe* (1914), 373-6

Mitchell, James. 'Wild Goats on Jura'. Letter to *Country Life*, (24 October 1957)

Moorhouse, Sydney. 'The Wild Goats of Britain', *The Yorkshire Post*, (9 May 1964)

Morley, Elizabeth. 'An Ancient and Pure Breed of Goats in Britain', *British Goat Society Monthly Journal*, 44 no 10 (1951), 174-5

171

Morley, Elizabeth. 'The Bagot Goats', *British Goat Society Year Book* (1955), 5-7

Mulligan, M. J. 'Goat Superstitions'. Letter to *The Field,* (29 March 1952)

Murray, Margaret A. *The God of the Witches* (1956), 43-4

Naismith, W. M. *The Islands of Scotland (excluding Skye)* *(Edinburgh* 1934)

Newell, J. P. 'Wild Goats Extinct in Co Galway', *Ir Nat J,* 13 no 4 (1959)

New Statistical Account of Scotland. Fifteen volumes (Edinburgh 1845)

Niall, Ian. 'Wild Goats', *Country Life,* (24 January 1957)

Nimrod. *Sporting* (1838)

Oliver, P. J. 'Goats in the Stable'. Letter to *The Field,* (24 March 1955)

Oswald, Arthur. 'The Wild Goats of Bagot's Park'. Letter to *Country Life,* (25 November 1954)

Palmer, W. T. Letter to *Country Life,* (28 March 1914)

Payne-Gallwey, Ralph. *The Fowler in Ireland* (1882), 323-4

Pearce, Jane. ' "Wild" Goats of Wales'. Letter to *Country Life,* (21 February 1957)

Pegler, H. S. H. 'Angoras, Toggenburg and Nubian Goats in England', *British Goat Society Year Book,* (1923), 81-5

Pegler, H. S. H. *The Book of the Goat* (6th edn c1925)

Pennant, T. *Tours in Wales,* edited by John Rhys. (Caernarvon 1883)

Pennant, T. *A Tour in Scotland and Voyage to the Hebrides* (Chester 1772)

Pennant, T. *British Zoology* (4th edn, Warrington 1776)

Pennant, T. *History of Quadrupeds,* 1 (3rd edn 1793)

Pliny, Caius. *The Natural History of Pliny.* Translated by John Bostock and H. T. Riley (1890)

Plowman, D. 'Wild Goats or Film Extras'. Letter to *Animals,* (August 1969)

Porta. *De Humana Physiognomonica.* Quoted by F. E. Hulme, (see above)

Preston-Tewart, A. 'On Getting One's Goat'. Letter to *The Field,* (26 February 1959)

Pycraft, W. P. *The Courtship of Animals* (N.D.)

Pye, Henry James. *The Sportsman's Dictionary* (5th edn, 1807)

Richmond, W. K. Wild Goats. *Scottish Field*, (May 1955)

Ridler, J. K. 'Goats' Milk for Calves'. Letter to *The Field*, (11 November 1954)

Ritchie, James. *The Influence of Man on Animal Life in Scotland* (1920)

Ritchie, James. *Beasts and Birds as Farm Pests* (Edinburgh 1931)

Robertson, W. B. 'Goat Superstitions'. Letter to *The Field*, (24 May 1952)

Robin, P. Ansell. *Animal Lore in English Literature* (1932)

Robinson, Louis. *Wild Traits in Tame Animals* (1897)

Robinson, Phil. *The Poet's Beasts* (1885)

Rodgers, John. *The English Woodland* (1941), 101

Ross, George. 'Are There Ibex in Ireland?'. Letter to *The Field*, (2 April 1953)

Ross, Winifred M. 'Wild Goats in the Highlands'. Letter to *Country Life*, (19 December 1947)

Rostron, G. E. 'Welsh Wild Goats'. Letter to *The Field*, (12 April 1956)

Royce, E. Letter to *The Field*, (10 May 1941)

Royds, Thomas Fletcher. *The Beasts, Birds and Bees of Virgil* (Oxford 1914)

Ruffle, The. 'Stalking the Wild Goat', *The Country Sportsman*, (1949)

Ruffle, The. *The Sporting Rifle in Britain* (1951), 67-71

Ruffle, The. 'The Camera Versus Wild Goats', *The Country Sportsman*, (July 1951)

Rutty, John. *An Essay Towards a Natural History of the County of Dublin* (Dublin 1772)

St John, Charles. *Sketches of the Wild Sports & Natural History of The Highlands* (1878), 222

Savonius, Moira. 'Legend of the Yule Buck'. *The Field*, (22 November 1952)

Schreiner, S. C. Cronwright. *The Angora Goat* (1898)

Schwarz, E. 'On Ibex and Wild Goats', *Ann Mag nat Hist*, 16 (1935), 433-7

Seager, H. W. *Natural History in Shakespeare's Time* (1896)
Shaw, C. Hamilton. 'Goat-Sheep Hybrids'. Letter to *The Field*, (4 November 1954)
Sheridan, J. R. 'After the Wild Goats and Birds on the Achil Cliffs', *Land & Water*, (8 April 1882)
Shiner, M. J. 'Wild Goats in Wales'. Letter to *The Field*, (6 October 1966)
Shirley, Evelyn P. *Some Account of English Deer Parks* (1867)
Simkin, R. I. 'Wild Goats in Snowdonia'. Letter to *Country Life*, (9 January 1948)
Sly, J. T. Letter to *Country Life*, (21 November 1947)
Smith, S. 'Wild Goats on Bute'. Letter to *Country Life*, (30 January 1948)
Stainton, H. 'The Anatomy of the Goat', *British Goat Society Year Book*, (1956), 43-51
Stewart, A. E. 'Wild Goats in Scotland', *The Field*, (16 December 1933)
Stewart, Major P. M. *Round the World with Rod and Rifle* (1924)
Stone, J. G. 'Goats of the Cheviots'. Letter to *Country Life*, (26 December 1963)
Strutt, J. G. *Sylva Britannica or Portraits of Forest Trees* (1822)
Tegner, Henry. 'Wild Goats of the Border', *Country Life*, (29 February 1952)
Tegner, Henry. 'Goat Superstitions'. Letter to *The Field*, (5 April 1952)
Tegner, Henry. 'Wild Goats'. Letter to *The Shooting Times*, (21 November 1953)
Tegner, Henry. 'The Cheviot Wild Goat'. Letter to *Country Life*, (14 April 1955)
Tegner, Henry. *A Border Country* (1955)
Tegner, Henry. 'Friendships Between Animals', *Country Life*, (12 September 1957)
Tegner, Henry. *Game for the Sporting Rifle* (1963)
Tegner, Henry. 'Wild Goats of Britain,' *Scottish Field*, (February 1965)

Tegner, Henry. 'The Lost Herd,' *Animals,* 6 no 3 (1965)

Tegner, Henry. 'Goat Stalking in Scotland,' *Shooting Times & Country Magazine,* (1 April 1967)

See also The Ruffle

Thornton, T. *A Sporting Tour* (1896)

'Timothy' 'Strange Companions'. Letter to *The Field,* (24 February 1955)

Topsell, Edward. *The History of Four-footed Beasts and Serpents* (1658)

Trenbath, R. 'Goat that liked Fireworks'. Letter to *Country Life,* (22 January 1959)

Tristram, H. B. *The Natural History of the Bible* (2nd edn, 1868), 88-97

Turbervile, George. *Noble Arte of Venerie or Hunting* (1576 reprinted 1908), 145

Turner, F. Newman. 'Goats for Marginal Acres', *The Farmer,* (Winter 1953)

Unwin, A. H. *Goat-Grazing and Forestry in Cyprus* (N.D.)

Vesey-Fitzgerald, Brian. *British Game* (1946), 198

'W'. Wild Goats of Cheviot, *Country Life,* (16 May 1908)

Wallace, A. R. *Island Life* (1892)

Wallace, H. F. 'The Chase of a Goat', *Country Life,* (10 March 1906)

Wallace, H. F. *Hunting Winds* (1949)

Wallace, Robert. *Farm Live Stock of Great Britain* (Edinburgh 1923)

Wallis, W. J. 'Goats and Cows'. Letter to *The Field,* (3 May 1952)

Ward, Rowland. *Records of Big Game* (1st edn 1892; 2nd edn 1896; 3rd edn 1899; 4th edn 1903; 5th edn 1907; 6th edn 1910; 7th edn 1914; 8th edn 1922; 9th edn 1928)

Warwick, B. L. & Berry, R. O. 'Inter-generic and intra-specific embryo transfers in sheep and goats'. *J. Hered,* 40 (1949), 297-306

Watkins-Pitchford, D. 'Wild Goats of the Welsh Mountains', *Country Life,* (21 November 1963)

Watney, S. 'Goats in the Stable'. Letter to *The Field,* (12 May 1955)

Watts, J. M. 'Goats' Milk for Calves'. Letter to *The Field,* (18 November 1954)

Weldon, E. 'Goat Shooting on Achil Cliffs', *Illustrated Sporting & Dramatic News,* (19 March 1881)

Wentworth Day, J. 'The Cashmere Goats of Windsor', *Country Life,* (14 October 1954)

Wentworth Day, J. *They Walk the Wild Places* (1956)

Whitaker, Joseph. *A Descriptive List of the Deer Parks and Paddocks of England* (1892)

Whitaker, T. D. *A History of the Original Parish of Whalley* (Blackburn 1800)

Whitehead, G. Kenneth. 'Wild Goats of Scotland', *The Field,* (10 March 1945)

Whitehead, G. Kenneth. 'The Horned Game of Great Britain', *Country Life,* (19 September 1947)

Whitehead, G. Kenneth. 'Chamois-Goat Hybrid', *Country Life,* (16 January 1948)

Whitehead, G. Kenneth. 'An Historic Herd of "Wild" Goats', *Country Life Annual,* (1952)

Whitehead, G. Kenneth. 'Cashmere Goats of Windsor'. Letter to *Country Life,* (11 November 1954)

Whitehead, G. Kenneth. 'Wild Goats of Wales', *Country Life,* (26 September 1957)

Whitehead, G. Kenneth. 'Wild Goats in Wales'. Letter to *The Field,* (20 October 1966)

Whytehead, J. L. 'Tassels', *British Goat Society Year Book,* (1926), 27

Williams-Ellis, John, 'Wild Goats in Wales'. Letter to *Country Life,* (5 December 1947)

Wilson, Samuel. *The Angora Goat* (Melbourne 1873)

Wrotteseley, George. *A History of the Family of Bagot* (1908), 35

Wyatt, Colin. 'Tree-Climbing Goats of Morocco', *Country Life,* (22 June 1951)

Zeuner, F. E. *A History of Domesticated Animals* (New York 1963)

ACKNOWLEDGEMENTS

Much of the information concerning the status and distribution of wild goats in Britain has been obtained through the co-operation of those owners and tenants of estates who have been kind enough to answer my enquiries. Their names are too numerous to mention but to all I would like to express my gratitude.

I also acknowledge with grateful thanks the help given to me by Miss M. F. Rigg, Secretary of the British Goat Society, who kindly made available to me some back numbers of the Society's *Year Book* and *Monthly Journal* in which reference to wild goats was made.

My thanks are also due to Messrs Geographia Ltd, for allowing me to use their outline maps to provide the background for my Goat Distribution maps.

Acknowledgement is also made to the editors of *Country Life* and *The Field* for allowing me to use some material which originally appeared as articles in these two papers.

Index

Brucellosis, 79
Buchanan Smith, A. D., 9, 158, 167
Buckingham, Duke of, 98
Buffon, Count de, 69, 71, 78, 167
Bullough, Lady, 91
Burren Hills, 52, 142, 163
Burton, Baroness, 158
Bute, Isle of, 45, 139, 159
Cairnsmore of Fleet, 46, 88, 96, 123, 124, 159, 160
Caley Park, 96
Calke Abbey Park, 96
Campbell, Mad Colin, 127
Capra, 13; *caucasica*, 18, 21; *falconeri*, 20; *falconeri cashmiriensis*, 20; *falconeri chialtanensis*, 20; *falconeri falconeri*, 20; *falconeri heptneri*, 21; *falconeri jerdoni*, 20; *falconeri megaceros*, 20; *falconeri ognevi*, 21; *hircus*, 13, 14; *hircus aegagrus*, 14, 22, 130; *hircus blythi*, 14; *hircus hircus*, 13; *ibex*, 14; *ibex ibex*, 16; *ibex nubiana*, 18, 130; *ibex severtzovi*, 18; *ibex sibirica*, 18; *prisca*, 23; *pyrenaica*, 14; *pyrenaica hispanica*, 15; *pyrenaica lusitanica*, 15; *pyrenaica pyrenaica*, 15; *pyrenaica victoriae*, 15; *walie*, 19
Carradale, 46, 116
Cartwright, Mrs., 24
Cashmere Goat: *see* Goat, Cashmere
cashmiriensis, Capra, 20
Caucasian bharal, 21
Caucascian ibex: *see* Ibex, Caucasian
caucasica, Capra, 18, 21
Caucasus, 13, 14, 18, 21, 23
Chamois, 48, 50; goat hybrid,

49-50
Chamois Hunting in the Mountains of Bavaria, 49
Champion, Major F. W., 159
Charlie, Prince, 120
Chevin, 96
Cheviot Hills, 24, 33, 40, 43, 67, 70, 108, 134, 161
Chialtan markhor: *see* Markhor, Chialtan
chialtanensis, Capra, 20
Circassian goat, 23
Coch y wden, 79
College Valley Foxhounds, 43
Colonsay, Isle of, 30, 136
Colquhoun, John, 54, 86, 117,167
Couturier, Dr, 49
Cox, N., 85
Crete, 86
Cwm, 55
Cyprus, 53
Dalimier, Paul, 51,168
Daniel, W. B., 10
Darling, Dr Fraser, 36, 141, 168
Dawson, Dr Christine O., 56
de Bertodano, B., 164
Deer, 13, 29, 40, 41, 46, 47, 56, 71, 79, 86, 96, 98, 99, 100, 104
de Neville, Hugh, 29
Dochfour, 55, 120, 160
Drabble, Phil, 100
Drum, The, 113
Dublin, 79
Dublin Museum, 144
Dunn, Dr A. M., 55
Eagle, 44
Easton Park, 96
Edward II, 102
Edwardes, D. J. W., 163
Edwards, Major B. M., 159
Egerton, Lord, 105
Egypt, 18, 69, 76
Emmanuel, Victor II, 16
Emmanuel, Victor III, 16

Epping Forest, 30
falconeri, Capra, 20
Fallow deer, 100
Field, The, 9, 60, 158, 177
Fletcher, R. G., 158
Foot and mouth disease, 110
Forestry Commission, 100, 109,
    115, 116, 117, 118, 119, 120,
    122, 123, 124, 125, 126, 127,
    129, 131, 132, 133, 134, 137
Fox, 43, 61, 62
France, 14, 16, 77, 97
Fraser, Sir Keith, 87, 130
Fulwood, Forest of, 29
Gabhar, 30
Galloway, 46, 123-124
Garbh, 30
George IV, 98
Glacier goat : *see* Goat,
    Schwarzhals
Glen Douglas, 87
Glentarroch Rocks, 43, 125
Goat : age of, 46; and cattle, 66,
    67; and horses, 67, 68; and
    sheep, 47-48, 67; and snakes,
    69, 70, 83; antipathy with
    deer, 29; as animal of sacrifice,
    74, 82; as animal of wealth,
    31, 74; as foster mother, 81;
    as weather prophet, 70;
    banned from ancient forests,
    29, 30; beard, 39, 59, 61;
    blindness in, 57; calmness in
    fire, 68; colour of, 36;
    dentition of, 45-46, 71;
    destructive to forestry, 30,
    51-54, 117; diseases of, 56, 57;
    domestic breeds : Anglo-
        Nubian, 27, 28, 140; Anglo-
        Nubian-Swiss, 29; Anglo-
        Nubian-Toggenburg, 29;
        Angora, 26, 77; in Britain,
        78;
    British Alpine, 28, 140;

British Goat, 29; British
Saanen, 25, 27; British
Toggenburg, 28;
Cashmere, 26, 77; in Britain,
    77, 96, 97-98, 111;
English, 23, 24, 25, 28, 140;
Irish, 24, 25, 28, 117, 130;
Nubian, 26, 27; Sannen,
    26, 27, 34, 48, 117, 136,
    140; Schwarzhals, 101,
    116, 130; Telemark, 35;
Toggenburg, 26, 27;
Welsh, 24, 25, 28;
drowning, 139; early breeding
    in wild, 41, 73; estimating
    age by horns, 45; estimating
    age by teeth, 45-46;
excessive lust of, 72-73,
feral, 31-35, 36, 37;
    dominance in herd, 38-39;
    habits of, 37-42, 90, 101;
    interbreeding with wild
    animals, 14, 48-50; nuisance
    to deer stalker, 33, 67; origin
    of, 22;
fighting, 39-41.
food of, 51-55; fondness for
    sea-weed, 43, 55;
gall bladder, 56; gestation of,
    41; goes deerstalking, 67;
    horn measurements, 45, 90,
    122, 159-164; horn utensils,
    79; horns of, 22, 37, 38,
    44-45, 49-50, 51, 71, 86, 90
    (*see also* Horns.); hung by
    horns, 29, 54, 82; *husbandry,*
    9, 171; importation from
    Ireland, 31, 32;
in fable, goat and fox, 61;
    goat and wolf, 61, 62;
injurious to plants and shrubs,
    52-55; intelligence of, 61;
introduced for stalking, 87;
kills child, 96;

180

Goat, *cont.*
legend, 58-73; of fugitive
saved by injured goat,
65-66; of merchant and
daughter's dreams, 62;
of Christmas goat, 59; of
Huw Murray, 30; of
procuring a favourable
wind, 61; of Robert Bruce
saved by wild goat, 65; of
how to catch Sargi, 60;
lore, 58-73, 80, 82-83; for
keeping rats away, 67; for
preventing contagious
abortion, 66; for quaint
method of breathing, 70, 71;
life saved by horns, 71, 86;
remover of swelled head,
66; sight at night, 71, 83;
smell removes infection, 66;
manure, 31; medicinal uses of,
31-32, 80-81, 82, 84;
milk, 24, 25, 28, 30, 31, 35, 47;
mortality of, 42-44, 55;
multihorned, 49-50, 51;
origin of, 9, 22-35; parasitic
worms in, 55-57; polled, 27,
38, 47, 132; poor swimmers,
35, 55, 139, 142; predators
of, 43, 44; recorded in place
names, 30, 115; rut of, 39-41,
57, 71, 101; sayings and
proverbs, 60, 66, 67, 68, 69,
70, 84; sheep hybrid, 47-48;
smell of, 33, 44, 57, 66, 67,
71-72, 91, 95; origin, 72;
tassels, 38; terminology, 10, 11;
tree climbing, 52;
use of beard, 39, 78; blood,
82, 83; cheese, 80; dung, 84;
excrement, 82; fat, 78, 79,
83; flesh, 78, 79, 83; gall, 82,
83, 84; hair, 76, 78, 83, 84;
hooves, 83; horns, 79, 82, 83;

liver, 82, 83; manure, 31;
meat, 79; milk, 79-82, 84;
mohair, 77-78; skin, 74-77;
spleen, 82; tallow, 79; urine,
66, 84; whey, 31, 32, 33, 80,
81; wool, 77, 83, 97;
value to crofters, 31; voice of,
41, 92; weights of, 46-47;
whey, 31, 32, 33, 80, 81;
white, 34, 87, 88, 97, 107,
116, 119, 120, 128, 133, 136,
137, 138, 139, 140, 141, 143,
160
Goats: wild herds in England,
107-110;
Ireland (Eire), 79, 141-145;
Northern Ireland, 137,
Scotland, 30, 114-141;
Wales, 47, 79, 105, 110-113;
Goats and forestry, 30
Goat-sucker, 69
Goddard, T. Russell, 158
Gordon, John, 114
Grecian ibex: *see* Ibex, Grecian
Grecian Islands, 13, 14
Green, Major A. E., 163
Greenland, 35
Greswell, W. H. P., 110, 169
Guadalupe, 53
Hafez, E. S., 38, 169
Hardwick, V., 160
Hare, C. E., 10
Hare, with horns, 50, 51
Hawaii, 53
Henry III, 29
Henshaw, J., 162
heptneri, Capra, 21
Hingston, R. W. G., 39, 170
hircus, Capra, 13
hispanica, Capra, 15
Holmes, G., 160
Horns, best heads, 45, 112,
159-164; four horns on goat /